Global Fishers Crisis

Global Fishers Crisis

Malcolm Albert

RANDOM PUBLICATIONS
NEW DELHI (INDIA)

Global Fishers Crisis

ISBN 978-93-5111-930-2

Published in 2016 in India by

RANDOM PUBLICATIONS

4376-A/4B, Gali Murari Lal, Ansari Road
New Delhi-110 002
Phone : +9111-43580356, 011-23289044, 011-43142548
e-mail: sales@randompublications.com,
info@randompublications.com, randomexports@gmail.com

Reprinted 2023

Type Setting by : Friends Media, Delhi-110089
Printed at : Replika Press Pvt. Ltd.

Preface

Global fisheries are under increasing threat from a combination of factors, such as high-tech harvesting which has led to overfishing, habitat destruction, and climate change.

The last thirty years have seen a huge increase in man's exploitation of the world's fisheries. As the annual catch has escalated far beyond sustainable levels in many areas, we are provided with yet another reminder that capitalism is a huge obstacle to any kind of rational planning.

'Over-fishing', 'discarded' fish thrown back into the sea by fishing fleets because they have exceeded their EU quota; illegal fishing by pirates in search of the expensive - and now rare - fish species, (at an estimated annual cost of £9 billion) and the continual threat to endangered species like blue-fin tuna are all consequences of what many have begun to call a 'crisis' in the fishing industry.

Over-capacity (too many fishing boats) leading to over-fishing, an over-reliance on destructive fishing techniques such as bottom trawling, and continued government subsidies for mechanized fisheries to the detriment of the more sustainable, employment generating non-mechanised (motorized and non-motorised) sector are the main causes of the current over-exploitation. This situation has been worsened by rampant pollution, destruction of breeding grounds such as mangroves and estuarine areas, hot water discharge from thermal power plants, industrial effluents, sewage from major urban centres and coastal over development.

Underpinning the 'crisis' is a wider story of social hardship; economic divisions; decline in traditional fishing communities and working practices, and environmental destruction. The driving forces of this crisis - namely the rapid expansion of global trade, the impact of the industrial food system and the dominance of supermarkets - have combined to put pressure on fish stocks, fishing communities and the health of the sea.

This book examines each of these reasons in detail and recommends an ecosystem approach to fisheries management to counter the impending crisis.

– Author

Contents

1

Fisheries Crisis in Development Countries

INTRODUCTION

Coral reefs around the world are in crisis. Over 25 per cent have already died or are severely damaged, and another 30 per cent are seriously threatened and may die--from global warming, overfishing, destructive fishing practices, and unsound coastal development. Contributing to the problem, many coral reef animals and products are collected globally for commercial purposes, including food fish, the marine aquarium trade, live food fish markets, curios and knick-knacks, jewelry, and traditional medicines. Global trade is leading to overexploitation of reef animals and the use of fishing practices that destroy the reefs. Recent surveys of reefs worldwide found that many species of high commercial value were absent or present in very low numbers, in almost all the reefs surveyed. Results suggest that almost all coral reefs have been affected by overfishing, and that there may be no pristine reefs left in the world.

The United States is the number one consumer of live corals and fish for the marine aquarium trade, and of coral skeletons and other dried animals for the curio and jewelry markets. American consumers are unknowingly contributing to the decline and degradation of coral reefs. As a major importer and world leader in both trade and coral reef conservation efforts, though, the U.S. can play a critical role in shaping conservation strategies, consumer awareness, and international trade policies.

The symposium, Global Trade and Consumer Choices: Coral Reefs in Crisis, held at the 2001 American Association for the Advancement of Science Annual Meeting and Science Innovation Exposition in San Francisco, California, featured five experts who spoke on the global trade in coral reef species and coral reef conservation. They analyzed the causes and consequences of global trade in coral reef animals, the role of the U.S. as a major importer, and presented solutions to reduce the negative impacts of trade on these unique ecosystems while promoting long-term benefits to local communities through sustainable resource use. This report presents their papers, as well as the following overview of their main points.

Franklin Moore and Barbara Best of the U.S. Agency for International Development note that "coral reefs are invaluable resources to local communities around the world, serving as sources of food, jobs and livelihoods, and as coastal protection." By one estimate, coral reefs provide economic goods and ecosystem services worth about $375 billion each year to millions of people. "However," observe Moore and Best, "international trade is driving overfishing and destructive fishing practices, such as the use of cyanide to collect live reef fish. The unsustainable and destructive use of these precious resources jeopardizes the potential of coral reefs to sustain local communities and future generations."

Roger McManus, former President of the Center for Marine Conservation and Senior Advisor for Oceans in the U.S. Department of the Interior, notes that the trade in coral reef species is in actuality trade in an entire ecosystem. McManus asks if this trade, though, is consequential in light of other threats posed by global change and pollution. Fishing and collecting of other coral reef species are probably not sustainable in most cases, are the most significant threat to many targeted species, and in many areas are a significant threat to the overall health of reef ecosystems. McManus adds "it is culturally and politically important to address all threats to natural resources to ensure equitable treatment."

McManus also argues that the U.S. should not allow the World Trade Organization to influence its actions in the global trade arena. "The U.S. should exercise its authority as an international leader in both trade and environmental conservation, and address its own trade policies on coral reef animals." He suggests that the Convention on the International Trade in Endangered Species of Wild Fauna and Flora (CITES) "was not designed to be implemented as a comprehensive wildlife management tool at this scope of [the coral reef] trade." McManus recommends the creation of more marine protected areas with no-take zones, as well as a phase-out of U.S. imports of wild coral reef species except where the trade can fulfill criteria for sustainability more stringent than those now in use under CITES.

Charles Birkeland, from the University of Hawaii's Hawaii Cooperative Fishery Research Unit, noting the diversity of coral reef ecosystems, the life history adaptations of reef species, and the close knit intricacies of species interactions, cautions that coral reefs are inherently more vulnerable to overuse than are other marine ecosystems. If the coral reef trade is to be sustainable, reefs must be managed using an ecosystem-based approach. However, the fisheries management laws of the U.S. constrain the use of best management practices, such as an ecosystem-based approach, and are particularly ill suited to promote sustainable management of coral reefs.

The Coral Reef Ecosystem Fishery Management Plan for the western Pacific, though, is charting new territory as the first ecosystem-based fishery management plan for U.S. waters. It adopts a precautionary approach that shifts the burden of proof through a permitting system, applies adaptive management through continuous use of new information, creates marine protected areas with no-take zones, and requires "insurance" against unforeseen ecosystem impacts.

There is already strong international concern that some coral reef species are threatened or may become threatened through trade. Those species are listed under CITES, and include 2000 species of hard (stony) corals, black coral, giant clams, Queen conch, and sea turtles. However, most of the coral reef animals in trade are not covered under CITES.

Susan Lieberman, former Chief of the U.S. Fish and Wildlife Service (USFWS) Division of Scientific Authority and currently Director of the Species Programme of WWF International,and John Field, also with USFWS, discuss the potential for international trade agreements, such as CITES, to address environmental concerns. They note that "under CITES, exporting countries are required to conduct scientific determinations that this trade is sustainable, and will not jeopardize species or their ecosystems." However, this ideal is often not fulfilled in reality. While the primary responsibility for ensuring the sustainability of harvests and exports under CITES rests with exporting countries, Lieberman and Field suggest that "as an importing country, [the U.S.] also must share some of the responsibility, whether through cooperative bilateral and multilateral efforts or regulatory actions." They add that it is "vital to couple those strategies with increasing public awareness... An informed consumer is powerful 'Äì and can guide industry best management practices, government policies, and of course Congressional interest and action."

Vikki Spruill and Lisa Dropkin, of SeaWeb, though, observe that "Americans have little knowledge of the real threats to coral reefs and the life they support." Surveys show that Americans believe pollution is the greatest ocean threat, and that most consider coral reef conservation a second tier concern. But, there is support for coral reef protection and there is growing momentum to use consumer choice to drive ocean conservation. Spruill and Dropkin note that "Americans often express their conservation values by what they purchase. We need to make Americans aware of how consumers can contribute to reef conservation by buying sustainably caught reef animals." They add that informal data suggests aquarium hobbyists want an industry based on sustainability and are willing to modify their purchasing behaviours to support reef conservation.

Clearly, there is an urgent need to address the trade threat to coral reefs. Dire as the picture may seem, the papers presented here point to a way forwards. Potential remedies are known. Some are already being tested 'Äì like coral reef reserves with no-take zones, and consumer awareness campaigns with likely application to reef species. To achieve an equitable and sustainable solution will require that all must be pursued, engaging communities, exporters and importers, governments, non-governmental organizations, consumers, scientists, and international institutions.

GLOBAL SOLUTIONS TO GLOBAL TRADE IMPACTS

Coral reefs are some of the most productive and diverse ecosystems on Earth. Although they occupy less than 0.25 percent of the marine environment, reefs support more than a quarter of all known fish species. Coral reefs are an example of a highly complex and interconnected ecosystem. The rich complexity found in the reef ecosystem is a result of the evolution of many symbiotic associations which couple photosynthesis with processes that increase the retention of limited nutrients within the system. Despite their high gross productivity, reefs rely on a fairly closed nutrient cycle that yields little surplus production for extraction. They serve as critical habitat for numerous tropical species, provide storm and wave protection for coastal communities, and yield renewable resources for human populations around the globe. Yet these same human populations, often in areas of rapid development and growth, are placing increasing stress on these vulnerable ecosystems, due to various economic and social pressures.

Coral reefs now face a multitude of anthropogenic stressors. According to a 1998 World Resources Institute report, nearly 58 per cent of the world's reefs are at risk from a range of human activities 'Äì global warming, marine pollution, coastal development, destructive fishing practices, overexploitation of resources, disease, runoff from improper land-use practices 'Äì and many have been degraded beyond recovery. The study concluded that overexploitation of resources, destructive fishing practices, and coastal development pose the greatest potential risks. The international trade in coral, reef fish, live rock, and other reef organisms also contribute significantly to the decline and degradation of reefs.

These stressors have a cumulative and synergistic effect on reef health, and often occur in countries that lack the institutional or financial capacity or political will to mitigate them. Thus, it falls upon the international community to continue to mobilize funding, foster cooperation, monitor trade, and provide expertise to coral range countries if the reefs are to recover and thrive in the 21st century. This chapter focuses on the role of international trade, and multilateral and domestic efforts to regulate and control that trade.

THE ROLE OF INTERNATIONAL TRADE

Coral reef species are traded as a variety of products in numerous international markets, including live reef food fish, traditional food fish, curios, traditional medicines, live marine ornamentals, coral and live rock for aquaria, limestone production, and construction materials. Each of these arenas has its own set of issues, management regimes, and industries but all are subjected to some degree of international control or scrutiny. The impact of international trade on coral reefs relative to other stressors (pollution, bleaching events, etc.) is not well quantified, but the fragility of reef ecosystems means that even secondary threats cannot be ignored. In fact, the partners within the International Coral Reef Initiative (ICRI) adopted a resolution in the fall of 1999 recognizing that "international trade in corals and coral reef species is contributing to the stresses on these systems."

The best available information on trade in coral species listed under the Convention on the International Trade in Endangered Species of Wild Fauna and Flora (CITES) is a recent study by The World Conservation Monitoring Center. Analysis of CITES records for black corals (from 1982-1997) and stony corals (from 1985-1997) revealed that 70 nations imported a total of 19,262 tons from 120 exporting nations over this period. The trade in dead coral peaked in the early 1990s but has since declined to levels comparable with the mid-1980s. The decline in trade is due to a number of factors including an export ban on dead coral from the Philippines, the primary source of coral in the 1970s. However, the quantity of corals traded live for the marine aquaria market has increased tenfold from 1985 to constitute more than half of the global trade in 1997, between 600 700 tons, and is continuing to increase 20-30 per cent each year.

The live food fish trade is centered in Hong Kong, which imports a variety of species from Southeast Asia and the South Pacific archipelagoes at a wholesale value of $500 million per year. Specimens imported to Hong Kong are also re-exported to China, but little or none of this trade currently affects the U.S.. The U.S., however, does import significant amounts of traditional food fish (*e.g.*, shrimp, spiny lobster, and queen conch) from coral reef ecosystems and thus has a responsibility to participate in their conservation as well.

The U.S. is a significant part of the curio and marine ornamentals industry: we import more than 80 per cent of the stony coral in trade, and more than 50 per cent of the marine aquarium fish. This trade is growing by 10-20 per cent per year and could have a substantial impact on reef health in the coming decade. In addition to a lack of science-based sustainable management, the marine ornamentals industry continues to experience destructive fishing practices (*e.g.*, sodium cyanide use), harvest of multiple age classes within species, and poor survivability of traded specimens. The magnitude and growth of the U.S. market in the marine ornamental industry over the last decade all indicate that the U.S. should take a leading role in promoting sustainable use of coral reefs within range countries.

Means to Address International Trade

The problems identified above are addressed in a number of arenas, including capacity building, cooperative management and planning, international regulation, and domestic U.S. measures. Each of these approaches addresses a slightly different aspect of international coral reef trade, and each has benefited from U.S. participation. This chapter will note all of these areas, but will focus on the regulation of international trade, in both a CITES and non-CITES context.

Capacity Building

The U.S. remains committed to building the capacity for marine resource conservation in developing nations, particularly those that are coral reef range countries. Tropical marine ecosystems are crucial to our development plans in food security, the environment, economic growth, health, disaster mitigation, biodiversity conservation, and climate change. Many U.S. government agencies have committed funds and personnel to this effort, including the Department of State, the U.S. Agency for International Development, U.S. Fish and Wildlife Service, National Marine Fisheries Service, and the National Oceanic and Atmospheric Administration. Agency programmes often focus on training or personnel loans to promote and implement coastal zone management, marine park administration, law enforcement, institutional decentralization, and fishery retraining. In addition, funds may be provided for partnerships with governments, non-governmental organizations, and academia.

International Cooperative Management

Since the mid-1990s, the U.S. has taken an increasingly active role in international fora aimed at coral reef conservation. These include the International Coral Reef Initiative (ICRI, an intergovernmental negotiation and planning forum), the Convention on Biological Diversity (which the U.S. has signed but not yet ratified; it includes active discussions over destructive fishing practices on coral reefs), the International Queen Conch Initiative, North American Wildlife Enforcement Group, World Customs Organization, ICPO-Interpol, Asian-Pacific Economic Cooperation, and the South Pacific Regional Environment Programme. Most of these fora have focused on means to identify and reduce destructive fishing practices in tropical systems (*e.g.*, sodium cyanide fishing), and the U.S. has participated actively through the commitment of funds, personnel, and expertise.

International Trade Regulation--CITES

The Convention on International Trade in Endangered Species of Wild Fauna and Flora (CITES) remains the primary instrument for monitoring and regulating trade in wildlife, including coral reef organisms. CITES is the only international legal mechanism with the mandate to protect species from overexploitation due to international trade. The opening paragraph of the Convention explicitly recognizes that each species of wild fauna and flora is an integral component of the natural ecosystem in which it is found, and it acknowledges that wild fauna and flora are to be conserved in perpetuity, in the wild. The ecological principles underlying this statement are reflected the treaty, which provides for its scientific underpinning.

There are now 154 countries that are Parties to CITES, making it one of the most extensive international agreements for species conservation and protection. CITES currently affects international trade in over 30,000 species of plants and animals. Over 2000 species of hard coral (all Scleractinia, organ pipe, fire, and blue coral), ten species of giant clams (Tridacnidae), and one conch species of CITES, in addition to several non-reef building coral species. However, no species of coral reef fish are listed under CITES.

CITES entered into force more than 25 years ago, and benefits global species conservation through required scientific and management findings and action, annual species-specific trade statistics, international oversight on range country resource management, and a system of dual control in exporting and importing countries.

CITES offers three different levels of protection and regulation for listed species. The species threatened with extinction, which are or may be affected by international trade; commercial trade in these species is prohibited. The species that may become threatened if their trade is not brought under control. Commercial trade species is subject to regulation, and is allowed only if permits are obtained stating that trade will not be detrimental to the species. The species that are listed unilaterally by range countries, which require international cooperation in monitoring trade; commercial trade is permitted. All coral reef species covered by CITES are currently, which places obligations on both exporting and importing Parties. The treaty requires that each signatory nation establish a CITES Management Authority (to issue permits and perform certain other duties) and a CITES Scientific Authority (to monitor biological sustainability of trade).

U.S. Industry Efforts

The U.S. marine ornamentals industry has recognized the need to promote a sustainable supply of organisms harvested in a non-destructive manner from source countries. Towards that end, the Marine Aquarium Council (MAC) has been established as a non-governmental organization (NGO) to start dialog between collectors, wholesalers, retailers, and hobbyists on these subjects and to develop a means to certify environmentally sound products in the worldwide marketplace. MAC intends to define, establish, and promote a certification scheme to be used throughout the chain of custody in the marine ornamentals trade, using best practice standards developed from multilateral consultations with industries, governments, and other NGO's. The standards developed within MAC could help alleviate the poor shipment survival of many marine ornamentals, and may attract more consumers to "environmentally friendly" products. Obstacles to overcome in this process include market economics (uncertified, cheaply collected specimens will cost less than certified ones) and industry consensus. Obviously, as an industry organization MAC's interest is to avoid additional regulatory action or any legislative restrictions on U.S. imports.

Domestic U.S. Measures

The USCRTF International Working Group appointed a small Trade Subgroup to examine the CEQ process and develop possible measures to channel U.S. demand towards sustainably harvested reef species. The subgroup recommended a variety of actions for the U.S. to consider, including:

1. The U.S. should prohibit domestic harvest or collection and the import or export of coral reef species and products listed under CITES that are not sustainably managed or from environmentally sound mariculture programmes. Exceptions might include organisms intended for approved captive breeding programmes, scientific research, or public display. Restrictions might be extended to other species of concern under certain circumstances.
2. The U.S. should work with members of the marine aquarium industry, environmental organizations and other stakeholders to develop, within a specified time limit, responsible practices and guidelines for collection and transport of coral reef species that reduce mortality rates throughout the trade stream, improve product quality, and ensure survival in captivity.
3. After working with stakeholders over a specified time period, the U.S. should prohibit the domestic harvest and collection of any coral reef species by defined destructive fishing practices, such as the use of reef dredging, explosives, or poisons.
4. After working with stakeholders over a specified time period, the U.S. should prohibit the import or export of any coral reef species unless accompanied by certification that the products were not taken through the use of destructive fishing practices.
5. The U.S. should develop a coordinated national strategy for conservation and sustainable management of coral reef species and ecosystems within the U.S., and then work with the international community to share this strategy and develop criteria for the conservation and sustainable management of coral reef ecosystems in other parts of the world.

All of these are important options to consider. Such efforts would be intended to channel the U.S. demand towards source countries with effective management plans and/or those species that are harvested sustainably.

The legislative option would create a legislative and regulatory situation comparable to the U.S. Wild Bird Conservation Act, which supports CITES by allowing importation into the U.S. of certain approved captive-bred species, but also allows for importation of wild-caught birds from approved science-based sustainable use management plans. It is a "reverse list" approach. The WBCA, passed in 1992, was stimulated by unsustainable trade, and challenges to effective CITES implementation by range countries, similar to what we are seeing now with the trade in CITES coral species. The WBCA also encouraged the rapid development of captive breeding and husbandry techniques analogous to those that are emerging in the marine ornamentals industry today. Such a legislative option for U.S. imports would certainly benefit from more discussion between Congress, the Administration, scientists, conservation organizations, industry, and others.

Options that relate to cooperative work with industry are already underway, working through MAC and others. Certainly best management and trade practices are vital to both ensuring sustainable trade, and to conservation of coral reef ecosystems. We personally believe that such "voluntary" efforts are often most successful when coupled with regulatory/legislative solutions and incentives, and as such further dialogue on these options would be very useful. Finally, all of these options address U.S. import (and export) policy options, and options available to the export and import industry. We believe that it is vital to couple those strategies with increasing public awareness. We strongly urge efforts to create informed consumers, so that anyone purchasing coral jewelry, coral reef-based curios, coral for a home aquarium, or ornamental fish, should be able to be aware of the origin of the products they are purchasing.

We are not recommending consumer boycotts of coral reef-based products. Rather, consumers should be educated to demand products that are produced in a biologically, culturally, and economically sustainable manner, and provide both economic benefits to local people, and incentives for coral reef ecosystem conservation. An informed consumer is powerful--and can guide industry best management practices, government policies, and of course Congressional interest and action.

THE GLOBAL FISHERIES CRISIS

Records kept by the United Nations Food and Agriculture Organization (FAO) also indicate that of the world's 15 main fishing regions, four are depleted and nine are declining. This global fisheries crisis is primarily a result of overharvesting. The world's marine catch has increased more than four times in the past 40 years--from 18.5 million tons in 1952 to 89 million tons in 1989, but that growth is at great cost to the environment, and ultimately, perhaps to world food security.

Several decades of overfishing in most of the world's major fisheries has pushed many commercially important fish populations into steep declines Canada's northern cod collapse, for example. Catches are falling, despite the fact that expanding fleets are fishing harder, spending more time, effort and money than ever before in trying to maintain them. Some commercially important stocks are in such a critical state that all fishing has been shut down, or sharply curtailed. Hundreds of millions of people traditionally dependent on fishing for food and livelihoods face resource depletion, competition from industrial and distant water fleets, and loss of access to traditional marine food supplies. The biological diversity of the world's oceans is threatend by the highly competitive race to catch enough fish to keep pace with rising international demand.

Rising demand and decreasing stocks have led to increased mechanization, over-capitalization of the industry and the buildup of excessive fishing fleets, particularly of the larger-scale fishing vessels that are responsible for the bulk of overfishing related problems worldwide. A modern 'factory' supertrawler can be longer than a football field and capable of catching and processing into various products up to 200 tons of fish daily. One of the world's biggest trawl nets could encircle more than a dozen "jumbo jet" Boeing 747 aircraft at its opening. Ships deploying such nets have a capture rate of about ten tons of fish per hour.

Since 1970, the world's fishing fleet has expanded twice as fast as world catches. The fishing fleet in China, the world's leading fish producer, is now around six times the size it was in 1979. As a result, excess fishing capacity has reached alarming proportions. There are about 3.5 million vessels boats currently fishing in the world's oceans. Intense competition between countries and rival fleets over access to fishing grounds has sparked numerous international disputes over fishing rights in recent years.

Fishing for Protein

Since 1989 the world~s annual catch has been dropping. Many argue that this decrease is an indication that fishing has oversteppped natures~s limists.

The FAO warns that the rising demand for fish and fish products, combined with shrinking global catches from declining stocks, will soon lead to the point where there will be a shortfall of fish for human consumption of more than 20 million tonnes each year. Average annual consumption of fish caught in marine and inland waters could fall from 10.2 kilograms per person in 1993 to somewhere between 5.1 and 7.6 kilograms by 2050. This threatens the one billion people, mostly in developing countries, who rely on fish as a principle source of protein.

Wasteful and Destructive Production

The increased fishing pressure and the competition amongst fishing nations and their fleets severely stresses fish stocks and the marine environment. The widespread use of unselective fishing gear and indiscriminate practices result in one-quarter of the all the fish brought on board fishing vessels each year being discarded, usually dead or dying, back to the sea.

Commercial fishing vessels throw back on average about 27 million tons of unwanted fish annually. That amounts to about half of all the fish caught from the oceans each year that are consumed directly by humans. Along with these, millions of other marine animals are being incidentally captured and killed in fishing operations.

Destructive Fishing Gear

Some fishing gear is particularly deadly for certain animals in some situations. Longline fishing boats kill many tens of thousands of albatross each year in southern hemisphere oceans. Driftnets indiscriminately killed millions of marine creatures, while targetting for just one or two commercially valuable species. Marine mammals are frequently killed in great numbers in trawls, set nets and purse seine nets. In addition there is severe damage caused by fishing operations that use destructive gear and fishing practices, like bottom trawling, that physically disturbs marine habitats such as the ocean floor, sea grass beds or coral reefs.

The Politics of Overfishing

Today, there isn't a fishing region in the world that does not suffer from fisheries management decisions designed to satisfy short-term economic or political objectives (or both) rather than protecting the marine environment and conserving fish populations. The bulk of the problems stem from twenty fishing countries whose fleets land 80 percent of the total marine catch worldwide.

1. China 2. Peru
3. Japan 4. Chile
5. USA 6. Russian Fed.
7. Thailand 8. Indonesia
9. Korea, Rep. 10. Norway
11. India 12. Iceland
13. Philippines 14. Korea, DPR
15. Denmark 16. Spain
17. Taiwan, Prov China 18. Canada
19. Mexico 20. Vietnam

Commercial fishing in most of these countries has been very poorly managed. Even in a few countries where relatively advanced fisheries management systems have been in place for many years they have, almost without exception, failed to control the conditions and stem the abuses that lead to overfishing and destructive environmental impacts. Indeed, in many countries, governments have played an important part in fueling the expansion of excessive fishing capacity and overexploitation by providing lucrative subsidies, taxpayer funded handouts. On a global scale, these destructive subsidies run up to $50 billion a year. The chronic failures of fisheries management in European waters of the northeast Atlantic and North Sea are particularly noteworthy since this area has probably the longest standing and greatest single concentration of fisheries research and management institutions in world. Yet, Europe's fish stocks are plagued by overfishing, and massive excess fishing capacity.

The member countries of the European Union regularly ignore scientific advice when setting their annual catch quotas: when scientists from the International Council for the Exploration of the Seas recommended a 40 percent cut in the 1995 hake catch to protect the stocks, EU fisheries ministers agreed to a mere five percent cut. In the EU, as elsewhere, it seems that a fishery must be proven to be on its death bed before any remedial action is taken.

In the United States a similar picture of failed fisheries management prevails. According to the National Marine Fisheries Service (NMFS), the federal agency overseeing U.S. fisheries, 80 percent of the known commercially valuable fish populations (122 aquatic species) are currently either fully or overexploited; 67 species of these, among them bluefin tuna, swordfish, oysters, hard-shell and soft-shell clams and red snapper--are overexploited. A further 79 species continue to be exploited even though the status of their populations is unknown. Little wonder why this is so. The country's eight regional fisheries management councils are primarily composed of those interests who benefit directly from increased catches: commercial fishers and other industry user groups.

It does say much for fisheries management regimes elsewhere in the world, if the European Union and the United States are considered to be amongst the leaders in the field of fisheries conservation and management. Quite simply, nature's limits have been breached by too many fishing vessels catching too many fish, very often in wasteful and destructive ways, and it cannot allowed to continue if the oceans and the human communities around the world that depend on them are to survive.

Governments and industry are responible for ensuring swift action to bring about urgently needed reforms, including substantial fleet reductions and substantial cutbacks in allowable catches, the elimination of destructive subsidies, major reductions in bycatch and waste, and more stringent environmental controls over fishing operations. The full utilization of available fish stocks and profit maximization for industry have been the key goals of short-sighted fisheries development, while protection for the environment has taken a back seat. This has proven to be the formula for disaster in fishery after fishery, the world over, with the disastrous consequences for marine ecosystems and humanity already plainly visible around the world.

EXPLANATIONS OF CONFLICTS

Ghanaian artisanal fisheries present some interesting dilemmas in terms of conflict analysis and common property resources. Although Ghanaian waters are legally open access, a powerful sense of culture and tradition, shared amongst all groups has maintained a common-property system in the face of an open access regime. There is a profound sense of history and cohesiveness in fishing communities along the Ghanaian coast and this has undoubtedly allowed all communities to be able to control who does and does not fish off the beach. The overwhelming sense in all villages that they were in control of their own destiny is probably also a factor that has allowed common property access rules to operate in these circumstances. The collective will to protect their right to the resource, and the fact that that collective will is recognised and respected by all the villages is probably the key to the success of the management systems here. Although game theory suggests that short-term gain will often win out over longer terms societal needs, this appears not be the case so far in Ghana. Those in the community willing to work towards social benefits still outnumber those looking for personal gain. Rising transaction costs are a significant issue in Ghanaian fisheries management-both at the local and national level. Increasing economic pressures are making the day-to-day life in the communities harder. There is a boom in building projects in the country-particularly related to the tourist industry in the coastal regions. So far, however, only two villages have reported land-based disputes and interestingly both of these are close to the rapidly developing tourist industry under one hour's drive from the capital.

In both cases a rush to secure prime spots of land has exposed the often fuzzy and complicated land rights in the region. While in one village the land dispute was a major issue but totally unconnected to fishing, in another the dispute between two rival villages (both claiming rights over a prime piece of real estate) has found its way out to sea. It was reported that whereas neighbouring villages would previously help each other at sea in the case of accidents, that is no longer the case and the incidence of damage to nets and canoes between the two communities is on the rise. Further research will identify how common a problem this is along the coast, but in the Greater Accra region it is would appear that externally generated conflicts are the ones that put the most pressure on the ability of the community to solve its problems.

Economic policies are also caused the price of inputs to rise. This has a number of knock-on effects in the community. The rising costs of inputs impacts upon conflicts over the price of fish as both buyers and sellers aim to maximise their profits in increasingly difficult circumstances. The cost of fishing rises which, whilst not immediately causing more conflicts, puts added pressure on the day to day existence of fishermen and brings economic pressure to bear on the rest of the community.

Despite the increasing costs of living around them the fisheries management institutions within the communities would appear strong enough to be able to rebuff external shocks.

The rise in transaction costs is also evident in other sectors. An interesting problem that was raised by a number of villages was the lack of landing facilities. One community in Accra reported that there were too many canoes attempting to land on a small beach, and the village of Mumford in Central region reported a similar problem. Here the fishermen are not able to beach their canoes and have two options: leave them out at sea or take them to nearby Takoradi. Given the nature of the sea the latter option is preferred which has a number of impacts on the village. Firstly the transaction costs of the fishermen rise because they are travelling further to land their catch and have to negotiate access with the relevant Chief at the other landing place. Secondly, less fish is consequently landed at Mumford which impacts upon the livelihoods of the women fishmongers there. Travelling to meet the canoes involves transaction costs that, with the current economic climate are rising daily. It also involves the less obvious transaction costs of negotiating with fishmongers at the new landing site who now have increased competition in the purchase of fish. Once again, it would appear that the strength of the village communities and the mutual respect amongst the neighbouring villages has led such incidences to be considered as minor disruptions to daily and, so far, are not the source of any destructive conflicts. The intrusion of trawlers into coastal waters is another factor that has seen transaction costs rise. Reporting the incident takes time and effort, resolving the issue requires even more. Travel and communication between many of the very isolated villages and the district headquarters is costly and time-consuming. The trawler owners are often based in ports far from the site of the incident and are reluctant to admit liability and often deny all charges. Despite mechanisms to resolve such conflicts, they often remain the most costly conflicts in terms of time and money. As the incidence of trawler incursions rises, so presumably will the transaction costs associated with this.

Only increased funding of a monitoring and enforcement capacity and negotiation with the trawler owners is going to be able to better deal with this problem. This is supposed to be covered under the remit of the FSCBP, but as it liable to be a long-term goal. The root cause of the trawler issue-declining catches and rising costs pushing the trawlers into illegal areas-has a number of long-term solutions.

However, the trawler sector is facing severe economic pressures and any measures are probably too painful to contemplate at this juncture and beyond the capacity of the State (economically and politically) to implement currently. The Community Based Fisheries Management Programme (CBFM) would appear to have had positive benefits to the communities. By using the existing institutional structure, the CBFM has enabled more formalised management to be introduced to communities without upsetting what was clearly a system that worked well to begin with. The only communities that appeared indifferent to the CBFM were those in urban areas (principally Accra). The reason given for this attitude is that fishermen in these communities tend to be better educated and have greater access to media and other information. They are more cynical of government promises of how things will get better in the future and more likely to rebuff any attempts by the government to interfere. Outside urban areas, the CBFM was the reason cited for the decrease in conflict or the satisfaction with the conflict management system in place. The gazetting of local by-laws was looked upon favourably by all those spoken to. So, in terms of transaction cost analysis, it could be argued that providing local institutions with a more 'legitimate' basis both within their own community and in the district as whole has helped maintain transaction costs at a stable level. In those communities where the incidence of conflict has declined, the CBFM may even have helped reduce transaction costs. There is, however, a corollary to this argument. Although the CBFM has had a number of positive benefits, it has also inserted bureaucratic systems into institutions that worked quite happily without them before. Further research needs to be done on this issue, but it would be interesting to see if CBFM has actually increased the transaction costs of some communities in terms of time spent at CBFM meetings, costs incurred in travel to meetings and the added costs of bureaucracy involved.

THE WORLD TRADE ORGANIZATION AND ENVIRONMENTAL CONTROLS

During that evening in Georgetown, I noted that the Clinton Administration would do nothing to curtail its role in this destruction. I reasoned that the Administration simply would not take on the World Trade Organization (WTO) and its policies supporting free trade over any other concern. State Department officials assured me that I was wrong. But, sadly, I was right, the U.S. would effectively do nothing to stem imports of coral reefs into our county and the international trade continues to grow virtually unabated. We have instead essentially limited our efforts to assisting exporting nations to care for their coral reefs, to education, and to appealing for voluntary restraint by consumers. Why can't the world's only remaining superpower curtail its domestic commercial imports of coral reef products when it is clear that the trade is unsustainable, destroying reefs, and undermining the present and future economies of developing countries?

Because the World Trade Organization won't let us. If Congress passed a law controlling trade in imports of coral reef products, one or more of the exporting countries would complain to the WTO that we are inappropriately interfering with free trade. The burden would be on the U.S. to prove to the WTO's satisfaction that our actions were justified. This may seem fair enough, except that the WTO has upheld every challenge to environmental controls, and the burden of proof is always on those seeking to help conserve and protect the environment and our natural resources.

If a country does not agree to curtail its "unfair" efforts to protect the environment, then the court provides for the offending country to suffer financial penalties. If the U.S. won't stand up to offended nations, consider what a small developing country would do in the face of threatening complaints from another nation bent on profiting from the elimination of offending pollution controls or wildlife protections?

WTO's unbridled advocacy of trade no matter what its social and environmental costs is unacceptable if we are going to foster the kind of natural resource stewardship that ironically will make sustainable economies possible. That is why worldwide opposition to the WTO is rising and won't go away.

Of course it is not necessary to open all environmental controls to attack to prevent some countries from trying to promote inappropriate protectionism. A solution to this problem must be forged if we are to continue the benefits of expanding world free trade, and protect our planet's environment and its wildlife.

The highest priority for the new U.S. Trade Representative, Robert Zoellick, is to secure "fast-track authority" from the Congress for the President to negotiate new trade agreements. He has criticized the arguably modest efforts by the Clinton Administration to ensure trade agreements do not undermine environmental protection. During his confirmation hearing, it was clear that Congress will be facing a contentious battle on how fast-track legislation will ensure needed environmental protection. One way to resolve conflict is to find equitable solutions.

TRADE IN THE LARGER CONTEXT

Well over a year ago I attended one of those Georgetown dinners in which policy makers and those that would influence them meet to come to political terms. The order of the evening was to convince environmental organization leaders to support Clinton Administration policies regarding free trade agreements and the World Trade Organization (WTO).

The coral reefs, not coral reef organisms, because as recent trade studies have documented, the trade is in dead and live coral, other coral reef invertebrates and associated algae, and fishes. This is literally trade in an ecosystem; the majority of the trade is in its underlying structure. (There is a conversation evolving referring to "biotic" reefs recognizing that a majority of reef components are not living corals, but including algae and other "reef" building invertebrates.) In any case, the largest importer by far is the U.S., primarily for the aquarium trade. The U.S. is a major importer, too, of sea horses and Queen conchs.

Is this trade consequential in light of global change, reefs being mined for building material, or otherwise destroyed by pollution? The answer is probably mixed. Global change is the major threat to reefs worldwide, and mining and pollution probably are the next biggest threats accounting for the largest losses.

Nevertheless, fishing and harvest for the aquarium trade are probably not sustainable in almost all circumstances, and in many areas may be the most significant threat to coral reefs. In cases where the trade is targeted at rare species, the trade may be a threat to their continued survival. Moreover, it is culturally and politically important to address all threats to natural resources to ensure equitable treatment. However one gives weight to the relative threats, the destruction of coral reefs by human action is clearly ongoing.

All of this continues despite the fact that:

- tourism is emerging as the number one world industry, and that coral reefs have enormous economic value as a sustainable basis of tourism for many countries, for fisheries, and for other purposes;
- many coral species are recognized as threatened by trade, or potentially so, and are protected by the Convention on International Trade in Endangered Species of Wild Fauna and Flora (CITES);
- there is clear acknowledgement that harvest of coral reef organisms is a significant contributing factor to the degradation of coral reefs;
- the U.S., recognizing the threat of trade to coral reefs, largely restricts the extraction of corals and coral reef organisms from most of its own federal, state and territorial waters; and
- many of the fishes taken for export for food and for the aquarium trade are captured using cyanide and other poisons, or other methods destructive to their habitats. The result is more often than not consumer fraud as diners and aquarium fanciers frequently buy fish that are in the process of dying from the poisons used to capture them. Many of the divers conducting the captures are exposed to the poisons, and are not trained divers, increasing incidents of death and injury.

TRADE IN CORAL REEFS BE SUSTAINABLE

Through Executive Order no. 13089 for the Protection of Coral Reefs, President Clinton directed the Secretary of State and the Administrator of the Agency for International Development, in cooperation with other members of the Coral Reef Task Force to assess the role of the U.S. in international trade in coral reef species, and to implement strategies and promote conservation worldwide to protect coral reefs. The report from the Trade Subgroup of the International Working Group to the U.S. Coral Reef Task Force helps to respond to this directive.

The conclusions presented above about the threats of trade to coral reef ecosystems echo those of the Working Group. The report noted an accumulating suite of references to coral reefs providing economic and environmental services worth hundreds of millions of dollars a year. The report suggests that almost all reefs have been affected by over-harvesting, and that there may be no pristine reefs left in the world. In addition the Working Group made numerous recommendations on how to address the problem, including establishing new authorities to reduce the adverse impacts of the trade.

There is a fundamental assumption in these recommendations that it is possible to have sustainable, commercial exploitation of coral reef ecosystems. Many coral reef species are slow growing and slow to mature. Reefs are generally characterized as ecosystems with large numbers of species with relatively few numbers of individuals. Whether one measures sustainability by sustained biomass production, biomass remaining from exploitation, or stability in biodiversity measures, many observers of coral reef ecology are skeptical that such exploitation can be conducted in a relatively benign manner. More likely it is through a series of no-take reserves in marine protected areas, and rotating exploitation in sacrifice zones, that the impacts of the exploitation can be managed. Such management regimes in Australia have been held out as successful examples of this approach. Whether such a regime could be held to be really sustainable would presumably be dependent on whether some steady state over some meaningful time frame could be documented.

Nevertheless, while more research is needed to ascertain the impacts of all threats to coral reef ecosystems, and to demonstrate sustainability of their use, the burden of proof now generally lies with those who would constrain exploitation. Current public process assumes that those who profit from these resources are conducting their activities in a sustainable manner notwithstanding our knowledge of coral reef biology and data on harvest methods, quantities, and impacts that belie such an assumption. That is not to suggest demonstrating confidence in sustainability would be straightforward. Thousands of species are involved in the trade, and the general tenor of the Working Group report is to seek sustainability on a species by species basis. No doubt, particularly with respect to live rock, many of the species in trade have not been described, the taxonomy of others is under debate or difficult to ascertain, and for most their biology is so imperfectly understood that our ability to suggest any level of exploitation as sustainable is speculative at best. Coupled with the fact that most organisms in this trade are identified in public documents at best to genera or even higher taxa, the conversation that meaningful decisions about the sustainability of harvest or trade is happening or could be so is highly questionable.

Imagine for a moment you are an official in Fiji in charge of issuing export permits under CITES. The species to be exported were taken legally and in a manner that was "not detrimental to the survival of the species." There are 2000 corals listed under CITES, and you are presented with an export application for "live rock," including some species controlled by CITES, others controlled by other domestic law, and some for which there is no domestic or international law for their management. What do you do? What is your decision? How did you come to it? How do you decide this shipment and the ones before it and after it can be considered sustainable uses of the resource?

Keep in mind that the major substantive CITES standard for exports species is that the export is "not detrimental to the survival of the species". There are many people involved in this debate that view this standard as equivalent to a requirement for sustainability, but a clear reading suggests the CITES standard relates to probable extinction rather than a capacity to maintain commercial exploitation 'Äì arguably two very distinct standards. As the Fijian official you are lucky with regard to this standard. All you really have to suggest to allow the shipment is that it is not likely to be detrimental to the survival of any species it contains.

However, the treaty also suggests that exports should be limited to maintain the species throughout its range at a level consistent with its role in the ecosystem. Whether this is a legal standard for issuance of the permit that is enforceable is unclear. Whether "consistent with its role in the ecosystem" is a standard equivalent to "sustainable" is also uncertain. Sustainability, defined or not, is not a standard of international trade law, and for those purposes you are free, as that Fijian official, for all intents and purposes to ignore the question.

Imagine now that you are a border official in the U.S., and you receive a shipment of stony corals from Indonesia with an official permit attesting that the material was legally obtained and taken in a manner not detrimental to the survival of the species. You have several problems. You may not have a clue what species are really in the container, and if you did, how many are controlled by CITES or the U.S. Endangered Species Act. Under current U.S. policy, the pressure is to accept the shipment unless you challenge the authenticity of the export permit. You could also question the underlying assertions concerning whether the specimens were legally obtained or in the judgement of the Indonesian officials whether the shipment was detrimental to the survival of the species included. You are likely not in a strong position to do this. In the experience of CITES implementation, you are also on very thin ice with respect to the "role in the ecosystem" standard. Hence, you approve the import and waive the shipment through. These two examples suggest that CITES simply was not designed to be implemented as a comprehensive wildlife management tool at this scope of trade.

Recommendations

The International Working Group of the Coral Reef Task force identifies a series of actions being taken by U.S. agencies and makes recommendations that inherently make sense and are or would no doubt be beneficial, including research, education, better enforcement of existing domestic and international controls, and enactment of needed, restrictive legislation. The legislation did not happen, in part because some in the conservation community did not want interference with their "sustainable" coral reef projects. Nevertheless, enforcement of existing law is not likely to alleviate significantly the harmful impacts of the trade, except in some highly managed habitats

I would emphasize and add to those recommendations the following with the view to strengthen their potential impact:

1. Establish more coral reef marine protected areas with no-take reserves. This is undoubtedly the most efficient means to conserve the abundance and diversity of coral reefs. The U.S. has demonstrated leadership in this regard with the designation of the Northwestern Hawaiian Islands Coral Reef Reserve, and Palymra Island and Kingman Reef National Wildlife Refuges.
2. United States leadership has been less admirable in trying to resolve the conflicts between promoting free trade and the conservation of natural resources. The U.S. government needs to be a leader in trying to find solutions.
3. The U.S. should phase out imports of wild coral reef species except in those specific cases where operations providing such commodities are operating in a manner demonstrated to fulfill some kind of numerically objective definition of sustainability. The current means and vision for controlling the trade are so disproportionate to its immensity as to render them largely ineffectual.
4. The U.S. government should reassess its role in dealing with global change, through its domestic management policies and international assistance, recognizing the impacts such change will have on coral reefs. One of the many consequences of sea level rise and global warming, will be the decline of coral reefs throughout much of their range.
5. All governments concerned with marine conservation should take efforts to reduce and discourage mining of coral reefs and to prevent otherwise their physical modification and pollution.

The 21st century will be the most important in the existence of coral reefs over the past 5,000 to 10,000 years. In this century, we will make important decisions either intentionally or by indecision on whether we want to conserve a majority of coral reef biological diversity and how many coral reefs we want to see survive into the next century.

With regard to the coral reef trade, what does it say about us as a species, when we continence destructive trade in one of our planet's richest and sensitive ecosystems? A trade that is argued excusable because it damages coral reefs less than other threats.

A trade that directly and predictably threatens the health and safety of our fellow humans. A trade that captures and holds other organisms for our aesthetic enjoyment when we know their lives are essentially treated as consumables being acquired largely by poisoning their bodies and maintained without expectation of their survival. How can we experience beauty and pleasure in such a process? What do we consider potentially sustainable about this exploitation?

COASTAL ECOSYSTEMS

By any measure, coral reefs are among the most diverse and valuable ecosystems on earth. Coral reefs occur in over 100 countries, most of them developing countries without the capacity or financial resources to adequately manage these vital resources. Reefs support at least a million described species of animals and plants, and another 8 million coral reef species are estimated to be as yet undiscovered.

According to one estimate, coral reefs provide goods and services worth about $375 billion each year-a staggering figure for an ecosystem which covers less than one percent of the earth's surface. Reef systems provide economic and environmental services to millions of people as shoreline protection from waves and storms, as places for recreation and tourism, and as sources of food, pharmaceuticals, livelihoods, and revenues.

In developing countries, coral reefs contribute about one-quarter of the total fish catch, providing food to an estimated one billion people in Asia alone. Globally, half a billion people are estimated to live within 100 kilometers of a coral reef and benefit from its production and protection. In light of expected climate change and associated sea level rises, coral reefs can offer a natural, self-building and self-repairing breakwater against wave and storm damage. These extremely valuable ecosystems constitute the economic base and future hope for sustained development in many countries, particularly small island nations.

Coral Reefs in Crisis

A recent report from the Global Coral Reef Monitoring Network estimates that 25 per cent of the world's reefs are already gone or severely damaged and that another third are degraded and threatened. This coral reef crisis is happening here at home in the U.S. and in far away places, in some of the most remote areas of the world.

Coral reefs are in serious trouble worldwide from a powerful combination of stresses that are threatening their survival, including:

- overexploitation of resources for subsistence and commercial fishing;
- destructive fishing practices that degrade and destroy the habitat itself;
- increasing coastal populations, which are expected to double in the next 50 years;

- poor land use practices and runoff of pollutants, sediments and nutrients;
- disease outbreaks, which may be associated with poor water quality and pollutants;
- coral bleaching, associated with increasing seawater temperatures and global change; and
- removal of coastal mangrove forests.

These direct and indirect human activities pose significant threats to coral reef ecosystems, and the human populations that depend on them, particularly small island developing countries. For example:

- In northern Jamaica, it is estimated that almost all of the reefs are dead or severely degraded from overfishing and coastal runoff. Fish stocks have declined to a point where local fishers are now straining fish larvae out of the sea for fish soup.
- In the Philippines, degraded reefs and fish populations have led to an 18 per cent decrease in the amount of protein in the average diet.
- Human impacts are also occurring on U.S. reefs, oftentimes for use as luxury items. For example, in Hawaii at Honaunau, the top ten aquarium fish species have decreased by 59 per cent over the last 20 years, and at Kona the most popular aquarium fish show declines in abundance from 38 to 57 per cent.

Even under ideal conditions, it would take more than a lifetime for some reefs to recover. We can no longer continue to take coral reefs or mangrove forests for granted, or to assume that they can support unlimited resource use or unmanaged global trade.

Trade, Mangrove Forests, and Coral Reefs

While coral bleaching may be one of the largest threats facing coral reefs, international trade is having significant impacts on even the most remote and pristine reefs. Recent surveys of reefs worldwide found that many species of high commercial value were absent, or present in very low numbers, in almost all the reefs surveyed. Results suggest that almost all coral reefs have been affected by overfishing, and that there may be no pristine reefs left in the world.

International trade is also posing significant threats to mangrove forests, another critical coastal ecosystem that is intimately connected to coral reefs. Mangrove forests serve as important nurseries for many reef species. They help to maintain coastal water quality by reducing the run-off of sediments, pollutants and excess nutrients from the land. Nutrients and energy flow between these habitats as species move between them.

In some areas of the world, the major loss of mangrove forests is due to the construction of shrimp mariculture ponds for the world market. The cheap shrimp we consume here in the U.S. comes with enormous ecological and social costs for the local communities where mariculture ponds are inappropriately sited and intensively farmed.

Trade Drives Destructive Fishing Practices

How does the international trade in wild coral reef animals and products more directly impact reefs? Primarily through overfishing and the use of destructive fishing practices. Live fish for both the food trade and marine ornamental trade are often caught with the use of cyanide or other poison, which temporarily stuns the fish for easy collection. Cyanide use is a serious threat to some of the world's richest coral reefs, as the cyanide kills corals and many other coral reef organisms. The lucrative and unregulated international trade in reef fishes drives the use of cyanide. It is estimated that since the 1960's, more than one million kilograms of cyanide has been squirted onto Philippine reefs alone, and the practice has spread throughout East Asia and the Indo-Pacific. Various explosives, such as dynamite and homemade bombs, are also used to kill fish for easy collection, but at an enormous cost to the reef which is reduced to rubble.

In Komodo National Park in Indonesia, about half of the reefs have already been destroyed through the use of explosives, forming beds of coral rubble that can extend several football fields in length. While the use of explosives to collect dead fish is usually for domestic trade, some of the fish that are only stunned will enter the international trade stream.

International trade is also driving the removal of the calcareous skeleton or base of the reef itself; reef skeletons are sold as "live rock" for marine aquaria. This base is the resulting accumulation of coral skeletons over tens to hundreds and thousands of years. Living coral, which constitutes the essential reef habitat for a myriad of species, is also collected and shipped live for marine aquaria, or killed and dried for the curio and shell trade.

Trade Drives Overfishing and Removal of Targeted Groups

In addition to destructive practices, international trade is driving overfishing and the selected removal of key groups from coral reefs. Major groups targeted for trade are:

- groupers and wrasses for the live food fish trade;
- dead fish and invertebrates for food, medicinal products, and ornamentals including sharks, sea cucumbers, sea stars, mollusks and sea horses;
- live fish, coral and other invertebrates for marine aquaria and the ornamental hobby; and
- "live rock" or the calcareous base of the reef for marine aquaria.

The marine ornamental trade for the pet industry often targets rare fish and coral species, as these can fetch the highest prices. The trade is also targeting large-polyped corals, which tend to be the slowest growing and the least common. By targeting the large groupers and wrasses, the live food fish trade removes key species from these ecosystems, thus altering their dynamics. The loss of some is comparable to the loss of major predators from terrestrial ecosystems. Other fishes feed on algae, and thus play an important role in ensuring that corals are not overgrown by more rapidly growing algae. The removal of coral for the marine aquarium trade and for use as curios and knickknacks, and the removal of the "live rock" base, reduces the essential reef habitat. There are strong economic incentives associated with this international trade. The live food fish trade through Hong Kong alone is estimated to have a retail value of about one billion dollars a year. Some species of fish, selected live from a restaurant tank, can sell for almost $300 per plate. The global retail of marine ornamental fishes and aquarium hobby supplies is estimated at $500 million. Last year, for example, a pair of rare fish sold for over $5,000 each. Over 1000 different species of coral reef animals are now traded for marine aquaria.

The impacts from international trade are quite different from other more chronic causes of reef degradation, as these impacts are felt even in the most remote, pristine reefs. The use of destructive fishing practices, such as the use of cyanide, is spreading throughout the Indo-Pacific as fishing boats venture farther to find new unexploited fishing grounds

There is already strong international concern that some coral reef species are threatened or may become threatened through trade. Those species are listed under the Convention on the International Trade in Endangered Species of Wild Fauna and Flora (CITES), and include 2000 species of hard (stony) corals, black coral, giant clams, Queen conch, and sea turtles.

CONSEQUENCES OF DESTRUCTIVE AND UNSUSTAINABLE FISHING PRACTICES

International trade is driving destructive fishing practices and unsustainable harvests from coral reef ecosystems, reducing the value of coral reefs to local communities and prospects for long term sustainable use. In some areas, depletion of stocks and the destruction of the reefs are threatening peoples'Äô food security. This international trade is a highly mobile trade; as stocks are depleted in one area or country, the trade moves on to other areas, often spreading the use of destructive fishing practices. Thus, the nature of the trade provides few incentives for long-term sustainable use by a community, and few benefits are channeled to the local communities.

The use of poisons and hooka gear can have serious consequences for the collectors themselves. Cyanide fishing poses human health risks to the fishers through exposure to the poison. A hooka rig is a low-tech approach to scuba diving that involves a compressor on the boat that pushes air down long tubes to divers below. Divers can spend many long hours under water collecting with hooka rigs. Unsafe diving practices by untrained divers can lead to the diver's "bends" and result in joint disease and even paralysis and death. Each week, several divers who have contracted the bends are taken by fishing boats in Honduras for treatment in the local diving decompression chamber. These divers have been collecting spiny lobsters to supply the growing U.S. appetite for seafood. Similar reports of injuries to divers come from South East Asian countries where hooka rigs are used for collecting marine ornamental fish and live food fish.

U.S. ROLE IN INTERNATIONAL TRADE

In 1998, in response to the coral reef crisis, the Executive Order for the Protection of Coral Reefs was signed. The Order created the U.S. Coral Reef Task Force as a way of coordinating federal and state efforts, and charges federal agencies with the conservation and sustainable use of reef resources both domestically and worldwide. The Task Force was also asked to analyze and address the U.S. role in the international trade of coral and coral reef species.

The results of the trade analysis reveal that while live reef fish for the food fish market primarily go to Asian markets, the U.S. is the number one consumer of live coral and marine fishes for the aquarium trade and of coral skeletons and precious corals for curios and jewelry. Inadvertently, American consumers are contributing to the worldwide decline and degradation of reefs.

A closer examination of the U.S. trade reveals that the U.S. was consistently the largest importer of live coral during the 1990s, importing over 80 per cent of the live coral and 95 per cent of the live "rock" or reef base. Ironically, the U.S. prohibits the collection of coral and live rock in its own waters as they are considered essential fish habitats.

In addition to coral, the United States imports nearly half (eight million) of the total worldwide trade in aquarium fishes (15-20 million/year). Many of the fish imported for the marine aquarium market in the U.S. are captured with the use of cyanide and other poisons, which kills non-target animals and the coral reef itself. Sustainability concerns will only increase with the growing international trade. The international trade in coral and live rock to supply the aquarium trade has increased at a rate of 12 to 30 per cent per year since 1990.

The U.S. Role in Addressing the Trade Threat

The U.S. is part of the problem. The U.S. needs to be part of the answer. As a major consumer and importer of coral reef organisms, a major player in the world trade arena, and a leader in coral reef conservation efforts, the U.S. has a critical responsibility to not only address the degradation and loss of coral reef ecosystems worldwide, but to also encourage more responsible trade. As consumers, the U.S. should discourage the use of destructive or unsustainable collection practices that may jeopardize the future potential of coral reefs to sustain the local communities who depend on them for food and livelihoods. Rather, we should reward and encourage responsible use of these precious resources, and shift the burden of proof of sustainable use, for commercial and recreational purposes, to the users.

We need to emphasize community-based management of coral reef resources so that people living on and around coral reefs may share in the profits from coral reef activities. Other exploiters, whose primary interest is in making money without sharing benefits with local communities, should not be allowed to profit from these precious resources.

Oftentimes, local communities or national fisheries departments lack the capacity to sustainably manage reef resources, or to resist the short-term, high gain, economic incentives associated with the live food fish and marine ornamental trades. The U.S. Agency for International Development (USAID) is the principal agency of the U.S. Government responsible for building capacity in sustainable resource use in developing countries. USAID is presently working with local communities and national governments in about 20 countries to assist them in conserving and managing their coral reef and coastal resources through capacity building for integrated coastal management, better land-use practices, sustainable fisheries management, and marine protected areas.

There are also immediate actions available to the U.S. public in terms of awareness and individual consumer choice. There is an urgent need to develop positive trade regimes so that only products from reefs under sustainable management plans are allowed into or out of the U.S., to ensure that consumer demand by Americans is not contributing to the decline and degradation of coral reefs.

We must change our view of how we treasure and value natural resources. For example, last month, the U.S. adopted new trade measures covering the import of antiquities from Italy into the U.S.; all antiquities from Italy must now be accompanied by documentation and certification has to how they were collected and where they are from, to ensure that they are from legitimate sources. We must take a similar approach to natural resources. The U.S. government is promoting the idea among other nations within the Asia Pacific Economic Cooperation forum that consuming nations must bear some of the responsibility for their imports, and it is considering this approach for reef resources.

The U.S. could play a significant role by helping reward responsible practices, creating market incentives for responsible behaviours, and requiring certification of non-destructive collection practices and demonstration of sustainable collection of coral reef species. In this way, government, consumers, hobbyists and industry members can work together to ensure a responsible trade.

CONSERVATION THROUGH CONSUMER ACTION

SETTING THE CONTEXT

To assess public attitudes regarding a variety of ocean issues, between 1996 and 2001, SeaWeb has worked with professional market research firms. The results reported here represent a subset of the opinions expressed by survey respondents. In all cases, the surveys were conducted using industry-standard market research techniques and included sample sizes that allowed for accurate calculation of statistical significance.

Most Americans care about the oceans and believe that the health of oceans is threatened. Fifty-nine per cent of those responding to a 1999 poll rated the overall health and quality of the oceans as "only fair" or "poor," while less than one-third thought the oceans are in "good" or "excellent" condition. The same poll showed that over half believe that the condition of the oceans is deteriorating, having gotten worse in recent years. An even larger percentage 'Äì nearly three-quarters 'Äì evaluate the condition of coastal waters as negative. Coastlines are where the majority of people interact most directly with oceans.

Studies consistently show that the public--caring most about human health--views pollution with greatest concern and as the topmost threat to ocean health.

- From among a range of environmental problems--water pollution, rain forest loss, development, overconsumption, etc.--people consider most urgent those with the most direct impact on human health, namely water pollution and toxic wastes.
- The 1999 poll gave people a list of specific ocean issues--ranging from oil spills, corporate farm runoff, and improperly treated sewage to coastal development, commercial overfishing, and damage from boating/jet skis--and asked respondents to rank them. Deterioration of coral reefs was a middle tier concern, ranked well below pollution concerns but above such concerns as damage from boating and recreational overfishing. While this places coral reefs in a context of other concerns, it is difficult to know whether people would have themselves volunteered reef loss in the survey.
- In 1996, while testing messages that would help communicate more effectively about oceans, SeaWeb found that people connect to the need for ocean conservation mostly through two themes--a sense of responsibility to future generations and concern for human health. Making the case for coral reef conservation could use examples that fit these themes.

Communicating about specific threats to coral reefs and solutions is complicated by public confusion about reef systems. Approximately four in ten Americans either believe that fish breeding grounds and coral reefs are found throughout the oceans (versus only in certain places, as is the case) or do not know.

Towards Coral Reef Conservation and Consumer Action

However, there is strong support for establishing marine reserves. Respondents in the 1999 poll included protection of ocean life and habitats--and coral reefs specifically--as some of the top goals for marine protected areas. These were placed well above such other goals as responsible management for oil exploration/drilling and commercial fishing, maintaining clean areas for swimming and diving, and providing recreational areas for boating. Nearly three-quarters of the public would support prohibiting the collection of tropical fish and corals from protected areas.

The public clearly is concerned about the state of the world's oceans. The public believes that humans can do lasting damage to the oceans, and supports efforts to strengthen ocean protection. However, many do not believe that their personal actions have much impact on ocean health. On the other hand, taking environmental action as a consumer--for example, recycling motor oil, cleaning up litter, eating only environmentally safe fish, and buying non-overfished seafood--is more popular than other kinds of individual action (such as contacting politicians and joining an environmental group).

Indeed, there are substantial numbers of people willing to modify their purchasing behaviours to help oceans. For example, many strongly support actions to protect oceans, even if it meant paying more for seafood. Sixty-two per cent of respondents in one survey said they would not eat fish classified by the government as overfished, and 44 per cent would only eat fish caught or farmed in a way that protects oceans.

Americans are increasingly connecting to conservation through consumerism and there is growing momentum to use consumer markets to drive ocean conservation.

- The Marine Stewardship Council has developed a certification programme for sustainable fisheries.
- Various organizations, like the Monterey Bay Aquarium, National Audubon Society, and the Chefs Collaborative have developed seafood-buying guides for consumers, chefs, and restaurateurs.
- SeaWeb has started the Seafood Choices Alliance as a trade association to provide the seafood sector with the information it needs to make sound choices and provide the best options for its customers.

For coral reefs, there are two areas of consumer concern--food fish and marine ornamental fish, coral and live rock. To address the consumption of fish, the Seafood Choices Alliance and other organizations are working to create a sustainable food fish industry. To address the trade in ornamentals, the Marine Aquarium Council seeks to create a sustainable trade through certification of marine ornamental fish.

In addition to formal opinion polling, SeaWeb conducted an informal survey of 77 aquarium hobbyists attending a conference of the Marine Aquarium

Societies of America. The results of this survey suggest that there is great interest in using purchasing behaviour to support reef stewardship. Most respondents indicated they want to support an industry based on quality and sustainability. Some hobbyists currently seek out suppliers of healthy animals, and are "very interested" in the source and collection methods of the fish they purchase. More importantly, they are willing to pay more for fish that are certified as being caught and handled in an environmentally responsible way. While based on a small, self-selected sample of hobbyists, these results point to a positive atmosphere for providing information that would allow marine aquarists to make environmentally sound choices.

Conclusion

There is a need to better understand the public's knowledge about and attitudes towards coral reefs so that campaigns and communications can be crafted more effectively. There is the opportunity to use consumer initiatives as a tool to augment legislative and enforcement solutions to the coral reef crisis. This will require communicating the link between reef conservation and informed consumer purchases--a difficult task in the face of public confusion about coral reef ecosystems and how they work.

However, saltwater aquarium hobbyists and the trade can be a force for change. And, with growing interest in linking consumer behaviours to conservation, there needs to be continued investigation of how environmentally responsible aquaculture--whether for food fish or ornamentals--can relieve pressure on coral reefs without unintended consequences for reef-based economies.

SUSTAINABLE DEVELOPMENT CONCEPT IN FISHERIES

In fisheries, development may be defined as "a process of change through which sustainable and equitable improvements are made to the quality of life for all or most members of a society". Adding the need for sustainability implies that these improvements need to be achieved without risk to the long-term stability of the ecosystem concerned.

In 1987, the World Commission on Environment and Development defined sustainable development simply as "Development that meets the needs of the present without compromising the ability of future generations to meet their own needs." This definition recognises that human needs can change with time, and that preservation of the integrity of the environment is necessary to ensure that future generations can realise the full potential of their own development needs. Another, more elaborate definition states that "Sustainable development is the management and conservation of the natural resource base, and the orientation of technological and institutional change in such a manner to ensure the attainment and continued satisfaction of human needs for present and future generations. Such development conserves land, water, plant and genetic resources, is environmentally non-degrading, technologically appropriate, economically viable and socially acceptable." This definition was developed by the FAO Committee on Fisheries in 1991, and acknowledges the need to achieve a workable balance between guaranteeing satisfaction of both present and future human needs including fulfilment of social and economic demands, and conserving the natural resource base. In response to the general goals identified, FAO developed a Code of Conduct for Responsible Fisheries that provides principles for ensuring sustainable exploitation of marine resources.

SEARCHING FOR OPTIMUM FISHERIES MANAGEMENT

The 1991 FAO definition noted earlier recognises the need for management of the resources concerned as one of the essential actions for achieving sustainable development. To accommodate the other components of the FAO definition, it is clear that such management must embrace a range of objectives, incorporating biological, economic, and social elements. Knowledge of the status of the resources and the likely impact of fishing activities and other factors on these resources provide for informed and successful management of fisheries.

In the early days of fisheries management, scientists focused on the biological modelling of fish stocks (Graham, 1935; Beverton and Holt, 1957). Assuming that fish stocks respond predictably to moderate levels of exploitation, with a defined equilibrium state, fisheries scientists are able to calculate the maximum level of catch that would allow sustainable exploitation. This sustainable maximum catch is referred to as the Maximum Sustainable Yield (MSY), and continues to be a key reference point in present-day fisheries management.

MSY takes into account only the sustainability of the resource and so satisfies only a biological objective. In view of this, fisheries economists have argued that fishing is conducted as a business and that the economic benefits of the activity must not be ignored. This led to the development of another management reference point, referred to as Maximum Economic Yield (MEY). MEY is calculated based on optimising the difference between the cost of fishing and the income gained, and usually occurs at a lower fishing effort than the fishing effort occurring at MSY.

Subsequent to the development of MEY, social scientists have noted the importance of addressing social concerns, to ensure more equitable distribution of economic benefits. The social objective of Maximum Social Yield (MsocY) was then recognised, and the associated reference point varies with the fishery situation.

These biological, economic and social management objectives, as currently defined, are now facing yet another challenge, with the development of chaos theory (Smith, 1990 in Symes, 1996; Wilson and Kleban, 1992 in Symes, 1996).

Chaos theory suggests that although nature is non-random, it is unpredictable. The equilibrium state of fish populations assumed by MSY is not accepted, and fish populations are believed to vary unpredictably within limits. Chaos theory therefore proposes 'ecologically adapted management', based on knowledge of the long-term stable ecological relationships, and requires more flexible management systems.

Achievement of Responsible Fishing

Sustainable fisheries development can be achieved through responsible fishing, which considers rational fishery management objectives that address a range of issues including the status of the resource, the health of the environment, post-harvest technology and trade, as well as other economic concerns, social benefits, legal and administrative support. In the case of shared resources, a co-ordinated approach to responsible fisheries management is essential, and Caddy and Griffiths (1995) proposed the following actions:

Regulate Fishing Effort

It is crucial to control fishing effort, and to avoid financial incentives that would contribute to excess fishing capacity. Excess fishing capacity and overcapitalization threaten the sustainability of the resource, as well as the industry.

Establish Code of Conduct for Responsible Fishing

This is needed to maximize benefits from the fishery, while avoiding wastage caused by indiscriminate fishing practices. In particular, it is important to reduce catches of undersized fish and non-target species, and to avoid use of gears that have a negative impact on the environment.

Establish and support regional/international fishery commissions and organisations concerned with management of shared resources.

The successful management of shared resources requires effective coordination by all the countries concerned. The relevant international fisheries agreements promote participation in, and financial support for, the work of these commissions and organisations.

Regular Consultation among Harvesting Countries

Parties sharing the resources need to consult and collaborate regularly so as to promote understanding and full cooperation. Set agreed management objectives and related reference points, incorporating a precautionary approach

Agreement on management reference points during the early stages of the fishery will help to ensure full cooperation of participants with management decisions. Where there is scientific uncertainty, a precautionary approach to management is recommended.

Develop Contingency Plans

Management plans should incorporate some contingency for dealing with sudden and unpredictable environmental changes caused by man-made or natural disasters.

Management should provide mechanisms for handling problems arising from resource user conflicts.

Protect Biodiversity

The biodiversity or species richness of an ecosystem is an important measure of ecosystem health. The preservation of biodiversity will ensure that present human development activities do not threaten the ability of future generations to meet their own needs.

Protect the Environment

There should be monitoring and control of waste disposal and pollution. In addition, every effort should be made to prevent discarding of entangling material that could trap and kill species or physically damage the environment.

Promotion of Research

Research should be conducted to support and inform various aspects of management. There should be fair and equitable distribution of benefits derived from the fishery.

Constraints to Caribbean Island States

For many small developing island states within the Caribbean, there are a number of constraints that pose significant challenges to the sustainable development of fisheries, including the following:

- *Fishing Practices:* Destructive fishing methods such as dynamiting fish are still practiced in certain areas. Additionally, fishers continue to use illegal mesh sizes that catch very small fish, and non-target species that may have little or zero market value.
- Inadequate knowledge of the resource and ecosystem
  Many Caribbean islands do not have sufficient detailed data to permit an accurate evaluation of the status of their resources and a good understanding of their marine ecosystem.
- Inadequate knowledge and recognition of social and economic conditions
  Many governments within the region still do not have an adequate appreciation of the social and economic potential of sustainable fisheries development, and hence still invest minimum resources in fisheries development and management activities.
- *Absence of Long-term Policies:* The lack of long-term policies makes it difficult for Caribbean countries to maintain pace with rapidly

evolving trends in global fisheries management approaches and trade strategies.

- *Capacity:* In many Caribbean countries, fisheries administrations are poorly staffed, and there are limited numbers of skilled and knowledgeable fisheries technicians, scientists and managers, and limited equipment and funds available for basic tasks such as data collection and research. There are few research institutions and regional organisations within the region, and some of these suffer similar limitations in available funding and a broad range of technical expertise.
- *Resource user Input:* In the Caribbean, fisherfolk organisations have not actively participated in the management process, mainly owing to a lack of good organisational and administrative skills. On the part of the fishers, this has resulted in a lack of trust in, and respect for, the governments that are responsible for making fisheries management decisions.
- Capability for monitoring, control and surveillance
  In many instances, there is little or no capacity for monitoring, control and surveillance activities, and no associated legislative framework.
- *Post Harvest Aspects:* Caribbean countries are just beginning to develop their export markets for fish and fish products, and face the challenge of satisfying stringent standards in all aspects of fish handling, processing and packaging, recently established by the importing countries.
- *Facilities:* Equipment and facilities are not upgraded regularly to deal with expanding fishing activities and increasing management demands.
- *Habitat Degradation and Pollution:* For some time, there have been signs that the Caribbean ecosystem is under stress. Degradation of habitat, particularly in the coastal areas has resulted from uncontrolled coastal development activities mostly associated with expanding tourist industries and overpopulation problems, as well as extensive sand mining and deforestation.
- *Financial Resources:* Many Caribbean countries are small island developing States with limited financial resources. This impacts negatively on their ability to cope with the wide range of issues required for successful sustainable fisheries development.

Conclusions

Caribbean governments need to appreciate the social, economic and financial potential of sustainable fisheries development. Fisheries administrations need to be given more financial resources, and properly staffed and equipped to address effectively and completely all aspects of sustainable fisheries development and management. Additionally, small island states should recognize the benefit of, and work towards, regional coordination in fisheries management activities, including sharing of expertise and resources for education, research, technology, monitoring, control and surveillance activities, and development of the relevant legislative framework.

Having removed constraints to responsible fishing, fisheries management must then strive to achieve a workable balance between defined objectives protecting the resource and its environment, and those seeking fair and equitable distribution of viable economic benefits.

2

Financial Crisis and the Fishery Sector

Turbulence in world financial markets is being transmitted to the fishery sector. Credit is getting scarce and margins, which already are comparatively low in this industry, are getting thinner. Export finance is also becoming more difficult to obtain with banks tightening up the conditions for issuance of letters of credit. With almost 40 per cent of all fish produced, whether from capture or aquaculture, now entering international trade, this could have a drastic effect on trade over the next months.

Likewise, the difficulties of many banks heavily involved in the financing of world capture fisheries and aquaculture development, such as the three Icelandic banks now taken over by the Icelandic state, are also limiting credit availability to the sector. Contrary to the fuel crises, which hit the capture sector particularly hard, the ongoing credit crunch is having a severe impact also on the aquaculture industry, and especially on the farmers producing carnivorous species, which are heavily dependent upon industrial feed. For these species, the production cycle typically lasts a couple of years before the fish reaches market size, making feed and finance the largest cost factors in production. On the demand side, consumers are reducing their discretionary spending and the decline in restaurants visits is hurting the fishery sector.

TRADE

Global trade in fish products is forecast to reach 54.5million tonnes in 2008, marginally below the previous year's estimate. Shrimp trade volumes are expected to fall over the next quarters, as poor demand prospects are encouraging importers to cut back imports and to draw, instead, from their frozen inventories. At the same time, tighter trade finance and export credits are encouraging producers and exporters to lean more on domestic markets. Despite poor domestic demand and increased landing prices in the sashimi tuna market, Japan's imports of tuna, including loins, increased by 4percent in January-June 2008, against the same period last year, but the growth may not be sustained. The United States' tuna market was negatively affected by weak demand for canned tuna and tuna pouches. The European Union import demand is also forecast to be depressed by rising prices of raw material. With weakening import demand in the three major markets, exports from Thailand, the largest canned producer, may decline, which is prompting the country to actively look for new markets in the Near East and in Mexico.

During the first half of 2008, imports of groundfish by the United States fell by 9percent, largely reflecting reduced purchases of both fillets and blocks, a situation likely to also prevail over the rest of the year. The United States usually exports Alaska pollack to China for processing and subsequently re-imports the processed product this year. Lower United States catches of Alaska pollack limited availability of raw material for the Chinese filleting and processing industry for re-export. Similarly, supplies from Northern Europe to China for processing was reported to face difficulty in securing adequate trade finance for exports. This could contribute to lower the volume of groundfish trade in the next couple of quarters.

By contrast, all the major import markets of octopus have reported increases in deliveries so far this year. Japan imports in the first half of 2008 were up 15percent, while those by Italy, traditionally the second major octopus market, rose by 12percent. Spain, also a significant market, increased January-June imports by 20percent. On the export side, Viet Nam is consolidating its position as a major octopus exporter. The country has managed to enter the Italian baby octopus market, a product mainly utilised in the processing industry of marine salads, where it represents a convenient alternative to cuttlefish. Squid imports by Spain, the major market in Europe, reported record imports during the first half of 2008.

However, squid purchases by Italy and Japan are likely to decline in 2008. The two countries are holding large supplies, as both stepped up their imports in 2007, when international prices were particularly low. Squid exports from Peru and the United States are forecast to expand, while a sharp decline in Indian and Peruvian catches may affect negatively their shipments.

Imports of farmed salmon in 2008 have held up remarkably well, especially in the European Union and Japan, but also in several Latin American countries. Purchases by the United States showed some weakness in the first half of 2008. World demand is still good for this species although supply problems in Chile could cause somewhat higher prices next year. Similarly, trade in tilapia is expected to increase in 2008, despite a slowdown in the second quarter of the year. Much of the expected expansion would be on account of larger purchases by Mexico and the Russian Federation, which would compensate for reduced deliveries to the United States. The brisk world demand is anticipated to sustain deliveries from non-traditional exporters in Latin America, as supply problems in China, arising from damage to the tilapia production centres incurred last winter, will limit the country's ability step up its exports.

Trade in catfish could also be boosted byincreasing demand from the Russian Federation, Ukraine and countries in the near East. Imports by the

United States and countries in the European Union, in particular Spain, are also expected to grow. The strong import demand is expected to boost exports by Viet Nam to a new record level. Trade in fishmeal, on the other hand, may be constrained this year by a poor import demand by both China and the European Union. Imports to Germany, in particular, were 30percent less over the first half of the year than in the same period of 2007. The decline in world import demand is likely to affect mostly deliveries from Peru. Weaker prices for several farmed species may also soften import demand in fish oil, much of which is used as a feed component in the aquaculture industry.

World Fish Market at a Glance

	2006	2007	2008	Change: 2008 over
			estim.	2007
			million tonnes	%
World Balance				
Production	138.0	142.6	144.2	1.1
Capture fisheries	89.6	91.8	91.0	–0.8
Aquaculture	48.4	50.8	53.2	4.7
Trade value (exports billion USD)	85.9	92.7	98.8	6.6
Trade volume (live weight)	53.5	55.0	54.5	–0.9
Total Utilisation				
Food	110.4	112.3	114.5	1.9
Feed	20.9	20.8	20.0	–3.8
Other uses	6.7	9.5	9.7	2.6
Supply and Demand Indicators				
Per caput food consumption:				
Food fish (kg/year)	16.7	16.8	17.0	0.7
From capture fisheries(kg/year)	9.4	9.2	9.1	–1.5
From aquaculture (kg/year)	7.3	7.6	7.9	03.4

UTILISATION

Total fish utilisation is expected to increase, sustained by a 2 per cent anticipated expansion in food consumption, which is forecast to reach 115 million tonnes in 2008. As a result, per caput fish consumption is estimated to hover around 17 kg in 2008, up from 16.8 kg in 2007. By contrast, utilisation as feed is set to decline by almost 4 per cent to 20 million tonnes, driven by a contraction in fishmeal production.

With credit becoming a problem for many operators, trade volumes and utilisation could suffer with declining imports. This may result in stagnating fish consumption in several of the major markets that rely on imports to meet their needs. Stagnating or falling incomes may induce some consumers to reduce their fish intake and/or to shift towards cheaper fish products. The shift towards cheaper fish alternatives would negatively affect the higher priced products, in particular shrimps. Consumers may also respond to changes in relative prices by turning towards farmed fish species, in particular, Vietnamese catfish, European seabream and farmed salmon. Following the downward revision of Chinese production statistics, world estimates of consumption have also been adjusted downward. As a result and contrary to the previous conclusion, aquaculture has not yet overtaken capture fish as a source of fish for human consumption. However, this possibility cannot be ruled out to materialise in the near future, given current prospects for strong growth in aquaculture.

ROLE OF TIFAC IN FISHERIES SECTOR

Technology Information, Forecasting and Assessment Council (TIFAC) is an autonomous society under the Department of Science and Technology (DST), Govt. of India. Following the Technology Policy Statement of 1983 and the Technology Policy Implementation Committee recommendations, TIFAC was set up in 1988 for keeping technology watch on priority areas and to promote actions. A landmark achievement of TIFAC has been the first major national long-term technology forecasting exercise known as Technology Vision for India up to 2020. Major thrust of TIFAC activities includes undertaking selected projects in identified sectors among which Agriculture and Agro Food Processing forms a major component.

AGRO FOOD PROCESSING SECTOR

Under the Agro Food Processing sector, the fisheries sub sector was constituted under the chairmanship of Dr. S.A.H. Abidi, Member, Agricultural Scientists Recruitment Board (ASRB), Indian Council of Agriculture (ICAR), New Delhi, and few other national fisheries experts as members, with a view of transforming the cottage level fisheries industry, by:

(a) Establishing commercial level plants, with high replication potential, for processing/ producing quality fish products to capture enhanced domestic and foreign market and
(b) Creating job opportunities for local community utilizing the local resources through science and technology inputs.

This chapter reviews the success stories of TIFAC fisheries projects on development of value added fish products as well its future targets.

Socio-economic Development

To ensure rapid socio-economic development, application of science and technology should be the foremost priority. In this background, TIFAC has initiated various projects on the development of value added aquatic products in different geographical areas of our country with different women SHG's/ Cooperatives in collaboration with fisheries institutes as well as state fisheries departments; who are providing the technical and logistical support to the projects. The projects are selected and approved by the Fishery Panel based on the technical as well as economic feasibility of the projects, as recommended by the experts in the concerned areas. A Project Review and Monitoring Committee (PRMC), consisting of experts nominated by TIFAC, are constituted for each project for reviewing and monitoring the progress made by each project. The PRMC will review the progress relating to the project and provides technical advise from time to time at various stages of completion. The PRMC after closely monitoring the developments in the projects may provide mid course remedial actions, if required, to overcome the unforeseen obstacles.

COMMERCIAL LEVEL COTTAGE INDUSTRY

The objectives of the projects are to establish a commercial level cottage industry by organizing women fishers and to improve upon their socio-economic status; to improve the utilization of low value fish/ shellfish by value added product development; improvement in quality and quality testing of the product; and market linkage.

Fish Processing and Preservation

The projects have facilitated and efficiently transferred the technologies such as fish processing, preservation, packing, quality control and marketing as developed by the Fish Processing Laboratory of CIFE. This was done through training programmes. The training and exposure imparted during the programmes enabled the society to address various product quality related issues through the implementation of quality assurance techniques. The advantages of the technology adopted include superior quality, greater shelf life, and higher acceptability. Various value added fish products were prepared from low cost fishes. The products include:

1. *Prawn Pickles*: Pickling is one of the safest means of easy preservation of fish/ shell fish. Pickles are good appetizers and they add palatability to the starch based blend to taste Asian dishes, besides being highly nutritious. The technology involved is simple and can be imbibed by rural women with great ease. A new recipe has been developed, by CIFE, to improve the acceptability and the shelf life. Novel methods of preparations were used to get pickle with good eating quality.
2. *Fish Papad*: Papad is a direct product commonly preferred all over India as a side dish. Fish is incorporated in papad to make papad more nutritious and tastier. The flour of black gram pulse (Urad dal) is used as the major ingredient. Fish was further added to make papad more nutritious and tastier.
3. *Fish Chakli*: Piston or ram type extruder is used in different regions of India to prepare starch or pulse based fried snacks. Chakli (spiral)

is one such popular product in Maharashtra. Fish meat is incorporated in chakli to enhance its taste, flavor and nutritive value.

4. *Fish Sandwich*: Using minced fish meat a product was prepared, which could be spread between bread pieces to make fish sandwich.
5. *Fish Paste*: The product is prepared by using fish flesh (minced) with spices and preservatives; which can be used as a side dish along with rice or bread having a shelf life of three months.
6. *Fish Noodles*: Fish flesh is made in the form of paste and then mixed with refined flour to produce delicious fish noodles.
7. *Fish Puff*: By using twin screw extruder a directly eatable cooked product containing ingredients like fish flesh, rice and gram powder with salt and spices can be prepared with a shelf life of six months.

Small Scale Commercial Cottage Industry

The projects have successfully demonstrated the potential for processing low value fish efficiently and incorporating them in acceptable, nutritious and safe, novel and traditional products in a small scale commercial cottage industry.

The projects enabled the poor fisherwoman to organize and to form SHG's for attaining sustainable livelihood. Training programmes were useful in respect to give SHG's an insight into modalities of organizing product processing and marketing. The project helped to create awareness on the potential for value-addition and market diversification among a wide segment of the local people, thus facilitating the technology to get replicated in other parts of the country. The fishery resources have been utilized maximum leading to reduction of post harvest losses.

Due to the higher value of value added products, when compared to low cost fish has earned the women group a handsome profit thus improving their socio-economic status. Stringent quality checking by CIFE resulted in production of value added products with superior quality, thus enhancing the shelf life and profit. The projects facilitated closer interaction between local traders through visits to major markets and trade fair participation. Continued buyer-seller matching, and provision of market information assisted in market development. At the end of the project period, the societies have shown an improved access to major markets through expanding their value added product range.

Value Added Fish Products

The success of projects in the area of value added fish products from low cost fishes have prompted TIFAC to replicate it in other areas and also to initiate projects in other areas also. The major thrust areas which TIFAC is focusing for its future activities include breeding and culture of ornamental fishes, production of sheedal; and caviar and fillet production from trouts.

1. *Ornamental fish breeding and culture*: Ornamental fish keeping and its propagation has become an interesting activity for many, providing not only aesthetic pleasure but also financial openings. It has reached new heights in the recent past and has gained accelerated momentum, thus providing employment opportunities to many including unemployed youth, women and other unskilled masses. Indian waters possess a rich diversity of ornamental fishes with over 100 varieties of indigenous species, in addition to a similar number of exotic species that are bred in captivity. The strategies to be adopted for boosting commercial production of ornamental fishes includes technical support and dissemination of recent research developments in breeding and culture; formulation and preparation of nutritious feed; health management; water quality management; and training programme for creating scientifically and technically skilled man power
2. *Sheedal Production*: Sheedal is a form of traditionally fermented fish product prepared in the North-eastern region of India. Apart from its delicacy and food value, people of this region like sheedal very much because of its medicinal value for preventing stomach related disorders and malaria. The process of preparation varies from place to place and community to community. *Puntius spp.* is the major species of fish used for sheedal preparation. For sustainable production and uninterrupted marketing of sheedal the approach involves training, mass production, quality up-gradation, and market development strategies.
3. *Caviar and Fillet production from trout*: Caviar and fillets are premium products which have high demand and fetches greater value in the international market. Caviar is produced from the roe of trout and filleting provides quality trout flesh. Trouts are fresh water fishes seen in cooler climate and high altitudes. Jammu and Kashmir is blessed with trout resource and has been identified as the ideal place to launch such a project.

For initiating a project on caviar and fillet production, the major identified inputs includes, sufficient raw material, sophisticated processing equipments, trained man power to operate the project, and recent technologies and research developments.

Aquatic resource is a vital tool for transforming India into a nutritionally secured nation. Development of value added aquatic products can bring both improvement of the socioeconomic status as well nutritional security in our country. The efforts of TIFAC in this direction have shown fruitful from the encouraging results obtaining from various projects. By producing quality value added products and establishing the market, the SHG,s can help India to march towards prosperity as well as self sufficiency.

INTEGRATED FISHERIES PROJECT (IFP)

Established as Indo Norwegian Project during 1952 this was renamed as Integrated Fisheries Project (IFP) in 1972. The Project is engaged in the development of technologies for harvesting and post-harvesting of marine fish resources. It has a well established Fishery complex consisting of a fishing fleet, a modern mechanical workshop and Slipway to slip vessels up to a displacement weight of 250 tonnes, an Ice cum freezing Plant, a well equipped processing unit for processing and marketing of diversified value added fish products, a research and development laboratory to carry out quality analysis of raw fish, finished products handled by the Project and testing and developing suitable packages for different value added products and Life Raft Servicing Station. The Institute has a unit at Vizag also.

Mandate of the Project

(a) Technology development and transfer to beneficiaries consisting of rural fishermen community, small scale industries and export processing houses through consultancy and job work.
(b) Value added product development by way of process and product diversification from all varieties of fish including low value and unconventional species.
(c) Imparting training in the field of Post Harvest Technology, Refrigeration Technology, Quality control and Value added Products.
(d) Providing consultancy and training for rural development programmes like supporting local fish farmers, self-help groups of fisher community and fishermen's co-operative societies functioning and Panchayati Raj Institutions. And also to Women empowerment programmes in processing and establishment of SSI units to self help groups.
(e) Popularization and test marketing of value added products from all fish varieties including low value and unconventional species.
(f) Extension of the popularization and test marketing of value added products to new areas and developing market in all states in a phased manner with added attention to rural areas enthusing entrepreneurs to enter into seafood processing industry.
(g) Function as a nodal agency on all matters associated with quality assurance, standardization of fish products and safety of fish products.

FINANCIAL RISKS

In practice, every fisheries global trade transaction contains an element of financial risk. Purchasing, production and shipment all place a financial burden on the transaction that forces the seller to determine how alternative terms of payment would affect liquidity during its different phases until payment – and how this should be financed. And, if the deal is not settled as intended, an additional financial risk occurs. In the case of subcontractors, who do not share the risks of the transaction and are paid according to separate agreements, the risk increases accordingly and even more so should the seller have to offer a supplier credit for a shorter or longer period. When it comes to larger and more complex transactions, this financial risk aspect is even more obvious. One of the major problems for the seller could be to obtain bankable collateral for the increased need for finance and guarantees. Even after production and delivery, the seller could still be financially exposed in the case of unforeseen events and delays until final payment. Sometimes the interaction between the seller and the buyer can make it difficult to establish the exact cause for the delay in payment and there are then fewer chances for the seller to refer to a specific breach of contract on the part of the buyer.

On the other hand, if the seller has paid enough attention when drafting the sales contract, including the terms of payment, then it is more likely that any reason for delays will be possible to determine according to the clauses of the contract.

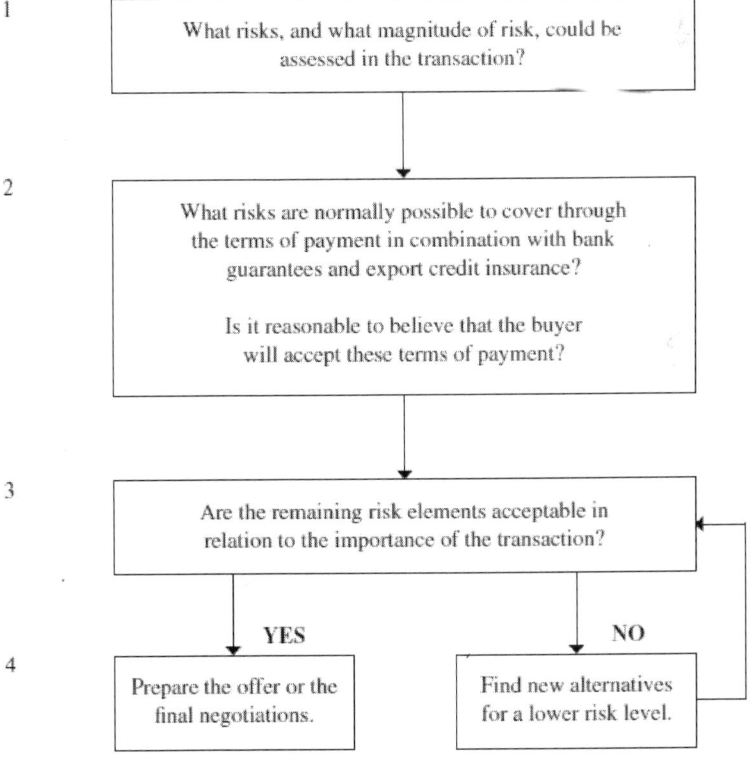

Fig. Risk assessment – a summary.

There could be numerous reasons for such delays, for example issuing a letter of credit too late, late changes in specification of the goods, late arrival of the vessel, congestion in port, changes in the transport route, to name a few. The real risk also tends to increase with longer and consequently more costly transport distances. Bureaucratic delays in many countries, as well as delays in the banking system, will have the same result – the final payment to the seller will not be made as anticipated according to the contract. Apart from ordinary overdrafts during production and delivery, the need for finance is also determined by the amount of credit that the seller may have to offer as part of the deal. If so, the financial risk is increased in line with the prolonged commercial and/or political risk.

A careful risk assessment is the first important step to a successfully completed transaction because it is the basis for the seller's own strategy, and for the final decision on what is acceptable in the negotiations with the buyer in order to minimize the risks involved.

FINANCIAL RISK AND CASH MANAGEMENT

Other forms of financial risk are more obvious but have to be underlined in this context; for example, if the seller misjudges the risks involved in the transaction and becomes exposed through terms of payment that do not cover the real risk situation, or mistakenly enters into the deal without proper risk protection. It goes without saying that such miscalculations can have serious financial consequences, from delays in payment to loss of capital. The financial risks are generally intimately connected to the structure of the terms of payment. The safer they can be made, the more the financial risk will automatically be reduced, the timing of the payments will be more accurate and the liquidity aspect of the transaction better assessed – in fact, the very essence of cash management. The safer the terms of payment the parties have agreed upon, the more costly they will normally be. And, if they contain bank security, such as a letter of credit or a bank guarantee, that will also reduce available credit limits within the buyer's own bank.

However, the buyer is often not prepared to accept higher costs and the use of their own credit limits in order to satisfy what might be seen as excessive demands from the seller, involving methods of payments, which in their opinion, are not normal practice in their country or normally accepted by the company. It is then up to the seller to evaluate the transaction, including potential competition from other suppliers. Eventually, the seller may have to accept the terms of payment offered and try to cover the remaining risks in some other way or to find a compromise by offering compensation to the buyer for the increased bank charges and/or the additional costs incurred by the use of the buyer's existing credit limits.

FISHERIES PRODUCT RISKS

Product risks are risks that the seller automatically has to accept as an integral part of their commitment. First, it is a matter of the product itself, or the agreed delivery; for example, specified performance warranties or agreed maintenance or service obligations. There are many examples of how new and unexpected working conditions in the buyer's country have led to reduced performance of the delivered goods.

It could be negligence concerning operating procedures or restrictions, careless treatment, lack of current maintenance, but also damage due to the climate or for environmental reasons. Matters of this nature may well lead to disputes between the parties after the contract has been signed and to increased cost for the delivery as a whole. It is important for the seller to have the contract, and specifically the terms of payment, worded in such a way that any such changes, which are directly or indirectly due to the actions of the buyer or originating within their country, will automatically include compensation or corresponding changes in the seller's commitments. This can be either in economic terms or in originally agreed time limits, or both. It goes without saying that these risks become even more complicated when it comes to whole projects or larger and more complex contracts. These are often completed over longer periods and involve many more possible combinations of interrelated commitments between the commercial parties, not only between the seller and the buyer, but also often involving other parties in the buyer's country, both commercial and political.

MANUFACTURING RISKS

The concept of product risk could also include some elements of the manufacturing process itself, even if in principle that subject falls beyond the scope of this handbook. This risk appears all too frequently when the product is tailor-made or has unique specifications. In these cases there is often no other readily available buyer if the transaction cannot be completed, in which case the seller has to carry the cost of any necessary readjustment, if that is even an option. Risks of this nature occur as early as the product planning phase but may often be difficult to cover from that time owing to the special nature of these products. But they also involve specific risks for the buyer, who often has to enter into payment obligations at an early stage but without the security of the product itself until it has been delivered and installed. In order to safeguard the interests of both parties, the terms of payment are often divided into part-payments related to the production and delivery phases, in combination with separate guarantees, to cover the risks as they occur in different phases of the transaction.

TRANSPORT RISKS AND CARGO INSURANCE

From a general risk perspective it is not only the product but also the physical movement of the goods from the seller to the buyer that has to be evaluated, based on aspects such as the nature of the product, size of delivery, the buyer and their country, and the actual transportation route. Most goods in fisheries global trade, apart from smaller and non-expensive deliveries, are covered by cargo insurance, providing cover against physical loss or damage whilst in transit, either by land, sea or air, or by a combination of these modes of transport. The cover under a cargo or marine cargo policy is almost always defined by standard policy wordings issued by the Institute of London Underwriters. These are called Institute Cargo Clauses. While there are numerous clauses that will apply to different cargos, the widest cover is provided under Institute Cargo Clauses A or with more restrictive cover under Institute Cargo Clauses B and Institute Cargo Clauses C. Cargo insurance is therefore normally provided through one of these Institute Cargo Clauses A, B or C, plus separate war clauses and strike clauses. This is also shown in the example of a letter of credit in the box on page 56. The question of who should arrange the insurance is determined by the agreed terms of delivery, as defined by the Incoterms 2000 and described earlier.

These terms also define the critical point during transport, where the risk is transferred from the seller to the buyer. That can be any given point between a named place at the seller's location and a named place at the buyer's location. That specified critical point determines the seller's and the buyer's responsibility to arrange insurance, as required. However, there is another aspect of risk coverage that the seller has to be particularly aware of, and that is the potential risk of the buyer arranging insurance according to some of the terms of delivery. If such a term of delivery is chosen, for example FOB and the buyer fails to insure in a proper and agreed way, the goods may arrive at the destination in a damaged condition and without adequate insurance cover.

If, at the same time, the terms of payment allow for payment after delivery, this risk de facto becomes a risk for the seller, who may end up with unpaid for, uninsured and damaged goods at the point of destination. Such a situation is obviously a consequence of the seller agreeing to terms of payment that did not cover the actual commercial risk, but the insurance risk involved could, in most cases, have been eliminated by separate seller's interest contingency insurance, as described below. From the seller's perspective, there are basically three different ways to insure the cargo, either with an open insurance policy covering most or all shipments within the seller's basic trade as agreed in advance with the insurer, or with a specific insurance policy, covering specific shipments on an ad hoc basis or those which are outside the set criteria of the open policy.

The open policy is by far the most common in fisheries global trade, normally reviewed on an annual basis, and with a 30–60-day cancellation clause, should conditions deteriorate substantially. The open cover is the most cost-effective alternative, but it also has obvious administrative advantages and will automatically secure the actual coverage of all individual shipments under the policy. The third basic form of cargo insurance is seller's interest contingency insurance, mentioned above, normally only offered as a complement to the open policy or as integral part of a specific policy, and on an undisclosed basis as far as the buyer is concerned. This insurance covers the risk that the goods may arrive at their destination in a damaged condition, resulting in the buyer's refusal to accept them or they may simply be unable or unwilling to pay for commercial or political reasons, including failure to produce a valid import licence. In such cases the insurance covers the physical loss of, or damage to, the goods, but it does not cover the credit risk on the buyer, which has to be covered through the terms of payment, in conjunction with any other arrangements.

The seller should bear in mind that cargo insurance is a specialized business, where cover and conditions may vary according to the commodity or goods to be shipped, the transportation route and the mode of transport, which is a major reason why open policy cover is the most common in fisheries global trade. But normal risk management procedures will always apply: new and adverse conditions and/or additional risks must be reported or approved by the insurer, and the policy normally excludes loss or damage due to wilful misconduct or insufficient, unsuitable or inadequate packing or container stowage by the assured party. Cargo insurance can be obtained directly from an insurance company or, very often today, directly through the transporting company or the forwarding agent handling the goods. In some countries it is also quite common to use independent cargo insurance brokers, who may be more able to select the most cost-efficient insurance package, based on specific conditions or the trade structure in each individual case. However, the seller should always ensure that the selected insurer has or is part of an established fisheries global network for dealing with claims and settlement procedures.

This is often also a requisite of the buyer, and if not explicitly agreed in the sales contract, such conditions may appear later on, for example in the insurance specifications in a letter of credit, as shown in the example in the box on page 56. More information about the cargo clauses and their coverage can be obtained through any broker, insurance or transportation company involved in fisheries global trade.

FISHERIES GLOBAL TRADE PRACTICES

All forms of business contain elements of risk, but when it comes to fisheries global trade, the risk profile enters a new dimension. Fisheries globally, you seldom have common laws that can support the transaction, as would be the case within one country. Instead, established trade practices and conventions are used to settle the undertakings made by the parties. The key to successful trade transactions, therefore, depends on a knowledge of these established practices and ensuring that the undertakings in the individual contract are in line with such practices. This is why it is crucial for the seller to have started with a correct risk assessment before finally entering into the transaction. Sometimes, however, the circumstances in a particular case are so obvious that one hardly thinks of it as a risk assessment, whereas in other situations a thorough risk assessment needs to be done. The main sources for fisheries global trade practices are publications issued by the Fisheries global Chamber of Commerce, which will be referred to many times throughout this book. In every new transaction one has to take it for granted that, from the outset, the parties will have different views about various aspects of the terms of payment. This is quite logical since the most important function of these terms for both seller and buyer is to minimize not only the risks involved, but also the cost of payment and of the financing of the transaction.

THE NEGOTIATION PROCESS

The seller will always try to get terms that will maximize the outcome and minimize the risk. However, they must also be prepared to accommodate reasonable demands from the buyer in order to match other competitors and reach a deal that is acceptable to both parties, thereby also developing a good long-term business relationship. Should the seller be inflexible on this point, it could result in an adverse competitive situation with the potential risk of losing the deal. On the other hand, demands from the buyer that are too stringent can have the same result, or be resolved by means of a higher price or some other amendment to the final agreement. The outcome of these negotiations will depend on past knowledge and experience, which is even more important if the buyer bases their request for tender on simplified or standardized terms of payment, usually to their own advantage. In many cases, such terms are adapted to conditions that are not optimal for the seller, compared with what the seller could have reached if they were individually negotiated. In such a case it is important to be able to argue and convince the buyer that there might be other solutions that can satisfy any reasonable demands, in order to find the optimal result for both parties.

There is, however, another – and in some countries very common – way to bridge the gap between the parties, if the seller has to abstain from some demands in negotiations with the buyer. The seller could approach a third party, often a credit insurance company, in order to reduce the commercial risk which could not be covered through the agreed terms of payment. Finally, it should be noted that the business practices which have been established over time in different countries or regions also create at least a common ground for both parties when starting their payment negotiations, ie choice of currency, form of payment and terms of financing. Local banks, trade councils and the chambers of commerce in both the seller's and the buyer's country can draw on their experience and give impartial advice on local business practice regarding both the form of payment and the more specific terms of payment, while also taking the size, commodity and other aspects of the potential transaction into account.

Such considerations can then be the starting point for negotiations between the parties.

DIFFERENT FORMS OF TRADE RISK

There are always potential drawbacks in trying to categorize such a general concept as trade risks which could have so many different forms and shapes, but it also has great illustrational advantages, particularly when they also coincide with commonly used business expressions. The main risk structure in fisheries global trade, which will affect both the seller's and the buyer's view of the terms of payment.

Obviously, all these risks combined do not often occur in one and the same transaction. For example, a sale to a Norwegian customer in USD may be just a matter of a straight commercial risk on the buyer, whereas delivery of a tailor-made machine to Indonesia has to be risk assessed in quite another way. In quite general terms, the risk structure is directly linked to the obligations undertaken by the seller.

This assessment can often be made relatively simple as a commercial risk only, but, in other cases, for example if the transaction also involves assembly, installation, testing or a maintenance responsibility, the assessment has to involve many other aspects as well. The question of risk is to a large degree a subjective evaluation, but it is still important for both parties to have a good knowledge of these matters in order to carry out a proper and meaningful risk assessment. Only thereafter does the question arise about how to cover these risks through the terms of payment together with other limitations in the contract, if applicable, and together with separate credit risk insurance or guarantees, as the case may be. It should also be noted that most export credit insurance, taken by the seller as additional security, could be impaired or even invalid should the seller themselves not have fulfilled – or been able to fulfil – their obligations according to the contract. This is another reason why it is so important that the obligations of the seller, according to the contract, are always directly related to those of the buyer. Otherwise the seller may end up in a risk situation that is worse than anticipated at the time of entering into the contract. When all the necessary evaluations have been done, the final decision as to whether the deal is secure enough to be entered into has to be taken. The worst that can happen is finding, after the contract has been signed, that it contains risks that the seller was unaware of at that time. It is then often too late to make changes.

TERMS OF DELIVERY AND TERMS OF PAYMENT

This handbook describes in detail the structure and design of the terms of payment as an integral part of the contract. However, the terms of delivery also have to be defined in order to determine when and where the seller has fulfilled the obligations to deliver according to the contract and what is needed in order to do so. There is a clear connection between these two sets of terms insofar as payment is mostly related to the point at which the risk passes from the seller to the buyer as specified by the terms of delivery; it is to be made either at that particular time or at a specific time thereafter. This connection makes it necessary to outline some basic facts about the different terms of delivery. The standard rules of reference for the interpretation of the most commonly used trade terms in fisheries global trade are Incoterms 2000 issued by the Fisheries global Chamber of Commerce Publication 560.

These rules are now generally recognized throughout the world, so any other unspecified trading terms, which may often have different meanings for companies in different countries, should be avoided. When agreed upon between the parties, these rules and their latest revisions must always be referred to in the sales contract and in all related documentation, for example CIF Hong Kong, Incoterms 2000. The basic purpose of these rules is to define how each Incoterm, as agreed in the sales contract, should be dealt with in terms of delivery, risks and costs, and specify the responsibility of the buyer and seller. For example, who should arrange and pay freight, other transport charges, insurance, duties and taxes?

These aspects are often referred to as the critical points in fisheries global trade, detailing at what point the risk is transferred from the seller to the buyer and how the costs involved should be split between the parties. There are presently 13 defined Incoterms, split into four groups, related to the seller's obligation to deliver the goods. Some Incoterms can be used only for maritime transport whereas others can be used for all modes of transport; similarly, some are more suitable than others for use in combination with terms of payment based on 'clean payments' in connection with open account trading, while others are used in combination with 'documentary payments'. These four groups are:

- Group E – where the seller has to make the goods available at their premises. The only example is EXW – Ex Works, where the seller must place the goods at the disposal of the buyer at the seller's premises or another named place not cleared for export and not loaded on any collecting vehicle.
- Group F – where the seller must deliver the goods to a carrier appointed by the buyer. For example, FOB – Free on Board.
- Group C – where the seller themselves must contract for the carriage of the goods, but without assuming risk of loss of, or damage to, the goods or additional costs due to events occurring after shipment. For example, CIF – Cost Insurance and Freight, where the seller delivers the goods when they pass the ship's rail in the port of shipment and must pay the costs and freight necessary to bring the goods to the named port of destination, but including also the procurement of

insurance against the buyer's risk of loss of, or damage to, the goods during carriage.

- Group D – where the seller also has to bear all costs and risks required to deliver the goods to the place of destination. For example, DDP – Delivered Duty Paid where the seller must deliver the goods to the buyer, cleared for import, and not unloaded at the named place of destination.

When choosing the appropriate terms of delivery, deciding factors include:

- the transportation route, the buyer and the nature of the goods, including the mode of transport;
- standard practice, if any, in the buyer's country or any regulation set by the authorities of that country to benefit their own transport or insurance industry;
- procedures, where the seller should avoid terms of delivery, which are dependent on obtaining import licences or clearance of goods to countries they cannot properly judge;
- the competitive situation, where the buyer often suggests their preferred terms of delivery and the seller has to evaluate these terms in relation to the risks involved.

For a standard delivery between established trade partners, neighbouring countries or countries belonging to a common trading area, this question is often easily agreed upon as a matter of standard practice with only an adjustment related to the actual freight and insurance charges, often in connection with open account trading. In these cases, the Incoterms Groups E and F are often used, where the buyer takes the main responsibility for transport and risk of the purchased goods. However, in other cases and when the seller wants to have better control of the delivery process and be able to select transport and/or insurance, the delivery terms C and D are more frequently used.

3

Biotechnology Development in Fisheries Sector

BIOTECHNOLOGY IN FISHERIES DEVELOPMENT

The gap between human demand and the availability of livestock products is huge in Africa, and imports have been increasing fast. The productivity of African breeds is very low: mature beef cattle of 4-5 years hardly weight 300 kg (as against 400 kg at one year for exotic animals); the best African cows produced 300 litre/lactation (as against 5 000 litres and more in Europe); 100 African ewes produced 50 lambs every two or three years (as against 150 lambs a year in developed countries).

Retrospectively, production policies that aimed at developing livestock through cross-breeding of exotic grade cattle with indigenous ones have generally failed. The new approach is not to make selection of local performant animals. Biotechnology offers good methods both for selection and cross-breeding schemes, essentially through Multiovulation Embryo Transfer (MOET) and the field diagnosis of pregnancy. Yet, there is wide diversity among the countries to use new technologies. For instance, even if simpler techniques such as Artificial Insemination (AI) are considered, it is seen that in East Southern Africa the technique is widely used, whereas in West and Central Africa, because of the predominance of nomadic transhumant livestock production, the technique is least developed.

MOET is well developed in few countries such as Zimbabwe and South Africa. Cattle farmers in these countries import frozen embryos from overseas. MOET is promising especially in the development of dairy cattle farms around large cities. Both AI and MOET are being used in Open Nucleus Breeding Systems (ONBS) to improve animal production. Small-scale ONBS projects are being carried out by FAO in small ruminants in Ghana and the Gambia. Moreover, by using semen from bulls, with genetic resistance to high temperature, diseases and insects common to African countries, embryos could be produced from superior cows of other regions and then implanted into surrogate mothers within the environment where improved livestock are needed. In animal feeding, biodegradation of low-quality forage is also a promising technology. It can be associated with chemical degradation techniques (use of urea and ammonia) to improve animal feeds during dry seasons.

In some of the countries, farm diagnosis of pregnancy is also being used by field workers by using a kit based on the difference of levels of progesterone in cow's milk. In the field of animal health, Sterile Insect Technology (SIT) is well known in Africa. Sterile males of glossina are being reared in Bobo-Dioulasso and Burkina Faso for field application. Successful field projects in tsetse control have been carried out in Nigeria (BICOT Project) and in Burkina Faso. Newworld screwworm was eradicated in the Libyan Arab Jamahiriya in 199091 by the use of SIT. In disease diagnosis, ELISA is being used to evaluate rinderpest immunity in vaccinated animals. Monoclonal Antibody Techniques (MAB) is not commonly used. The two techniques are very promising because they are very simple and most effective. Trials are under way to use the ELISA test in trypanosomiasis diagnosis. In vaccine production recombinant rinderpest vaccine trials done in the United States seem ready to be utilized in a mass vaccination campaign against rinderpest in Africa. There is an increasing level of activity in hybridoma work in sub-Saharan countries.

The work of ILRAD in Kenya is addressing the problem of diseases and parasites that affect cattle. Some of the common cattle diseases are East Coast Fever, tickborne diseases and trypanosomiasis. The focused approaches to researching on these diseases is expanding to yield results that will significantly benefit cattle breeders. The development of rinderpest vaccines remaining stable at ordinary temperatures was of high practical value.

Biotechnology in fisheries is recent in Africa and it is being applied essentially in private fisheries sector. Work is mostly confined to hormonal treatments for sex reversal and pawning, detection of fish and shrimps disease and development of fish feed.

RESPONSES TO THE CHALLENGES OF FISHERIES USE

The very reality that the root causes of the over-use of fish stocks are structural makes it nearly impossible to completely alleviate the pressures of rising demand, increasing abilities to pay for high-value species, and more effective ways to capture fish on an open-access resource. But people, organizations, and governments have responded to these realities in many ways, their responses sometimes improving the situation, but often exacerbating it. Two of the most prominent and important responses have been subsidies and management schemes.

The first, subsidies, occur when trade measures directly influence the income or reduce the costs of production for industry. Management schemes, the second, encompass the ways in which fisheries are governed and monitored.

In practice, the way this is done often highlights a lack of effective management. Subsidies and management, which can be seen as responses by individuals, groups, or governments to the root causes of fisheries depletion, are also the areas where the most opportunities for change in the way fisheries resources are used exist. Effective management regimes are considered by many to be key in the sustainable use of fisheries resources. Theoretically, effective management encourages a sustainable use of resources through providing conservation regulations and incentives for responsible use of the resource. They are especially beneficial under the pressures of international integration; responses by producers to the economic incentives from trade liberalization, for example, can be better regulated when effective management schemes exist so that over-exploitation and destruction are less likely to occur. Unfortunately, the prevalence of such management has been scant in real-world fisheries.

As the United Nations Environment Programme (UNEP) reports, "very few fisheries management systems have demonstrated the ability to keep catches below levels that put pressure on stocks." As noted in a study by the World Conservation Union (IUCN), "national and international progress on trade expansion and liberalization outpaces progress on fisheries management and articulation of sustainable development strategies." Fisheries management, crucial to sustainable use of the world's fisheries, has been difficult to implement effectively and has often left fisheries open to influences from other sources such as state trade measures.

Subsidies, one of these trade measures, have played a proven role in aggravating the causes of fisheries depletion. Theoretically, subsidies can lead to overuse of a resource by altering the behaviour of producers. Revenue-enhancing or cost-reducing subsidies, such as those that will be examined in this chapter, increase marginal profits at each level of effort and therefore theoretically lead to an increased overall effort. This may have the short-term effect of creating additional economic rent for producers, but either new entrants or increased effort by existing producers, also encouraged by the subsidy, will shift the level of effort so that rent is eventually dissipated. Assuming that the management programme does not effectively impose a sustainable level of resource use, cost-reducing and revenue-enhancing subsidies will push the level of overcapacity and overall effort even further than would an open-access resource in the absence of subsidies. The main theoretical effects of cost-reducing, rent-increasing subsidies are increased capacity, a delay in exit by producers from the sector, and increased effort by producers. These effects can cause the overuse of a resource, which can have many other significant impacts such as threats to food, livelihood, and resource security.

FISHING SUBSIDIES

In the real world of governments, fishers, fisheries resources, and the incentives that link them all together, a subsidy can take many forms. Fishing subsidies currently account for large portions of many industrialized country budgets; a 1998 World Bank study estimated annual budgeted expenditures on global fisheries subsidies to be between $14 and $20 billion, and that subsidies account for 20 to 25 per cent of the sector's revenues.

One type of subsidy, domestic assistance programmes, form a clear example of government endorsement of the fishing sector. These programmes are defined as budgeted assistance to fishing production, and often enhance operations and capacity. Japan, one of the biggest subsidizers of fishing activities in the world, has used domestic assistance programmes to recruit young fisherman, give aid to fish cooperatives and boat owners, support marketing, and encourage price stabilization. A 1998 World Bank report estimated that Japan budgets $270 million a year for these types of programmes, and a total of $750 when aid for fishing vessel insurance is included. The European Union (EU), another major subsidizer, supports such structural programmes as fleet renewal and modernization, processing and marketing, aquaculture, maintenance of port facilities, and genetic product promotion in their fisheries sector.

In addition, it pays for market supports to the sector in the form of a minimum import price programme and measures to support price floors. Costs of these programmes to the EU in 1996 were estimated to be about $530 million. While estimates of exactly what types of domestic subsidy programmes the Chinese government uses are hard to determine due to scant data, they have clearly aggressively promoted the expansion of their fishing sector in the last 15 years (and have become the world's leading producer) and are estimated to budget between $500 and $750 million annually for domestic subsidies. While other countries subsidize their domestic fishing industries, they have comparatively small budgets for such programmes as so do so on a smaller scale. The shear magnitude of the domestic subsidies in industrialized countries suggests the possible impacts on the behaviour of fishers their presence may cause.

A less obvious, but yet potentially just as influential, type of subsidy occurs when a country pays the access fees for their fleets to fish in foreign waters. Unlike domestic subsidies, these payments are not made directly to fishers or even domestic programmes, but to the foreign governments who have jurisdiction over the fish stocks. The EU is one of the major subsidizers of distant-water fishing through its numerous access agreements with developing countries, mostly in West Africa.

For example, the EU signed a five-year access agreement with Mauritania in 1996 that lifted an EU embargo on fishery imports from Mauritania and specified EU payments of almost $350 million over five years to the Mauritanian government. In return, the Mauritanian government granted increased access for EU fishing vessels in their waters, authorization of higher EU total harvests, and the specific allowance of EU-directed fishing of highly valued squid and octopus.

The monetary benefits to the "cash-strapped" nation were described as a "windfall;" in consequence, the number of eligible EU boats rose from 165 to 240 and allowable EU harvests from 76,050 to 183,392 tons. The EU has similar agreements with other West African nations, as well as others in East Africa, the Indian Ocean, and South America. Another distant-water fleet subsidizer, Japan, currently budgets about $200 million a year on distant water arrangements with foreign countries and foreign fisheries assistance programmes (which often help secure fishing rights). These payments went mostly to developing countries in the Pacific Ocean and elsewhere. The subsidization of distant water fleets from industrialized nations can have serious effects on the coastal communities of the developing countries with whom these agreements are typically made.

Possible Impacts of Perverse Subsidization

The impacts of government subsidies of fishing are at least as complex and numerous as the types of subsidies themselves, but in the context of sustainable development those related to market access due to trade measures and resource, livelihood, and food security are the most relevant. The trade implications of the subsidization of the fishing industry arise when the lower costs producers see as a result of the subsidies repress world prices for fish products. Those producers, who are often from developing countries that cannot afford subsidies, are then excluded from the market because they cannot afford to produce at the world price.

For example, when Japan subsidizes its fleets to fish in the waters of Chile, Japanese fishers bring the fish back to Japan and sell them on their domestic market. Chilean fishers try to enter the Japanese market and sell these same Chilean fish to Japanese consumers, but are kept out of the market because their costs are higher and the prices on the Japanese market are too low. In effect, Japanese fishers push Chilean producers out of the market for the sale of their own fish. This example illustrates the way in which subsidies in the fishing sector can exclude those who are outside the national subsidy regime from participating in the market.

Due to numerous examples where the presence of subsidies seemed to increase the capacity in the sector and, in turn, deplete available resources, fishing subsidies have also been blamed for a tendency to encourage overcapacity and overfishing.

The case of stock depletion in Canada's Northwest Atlantic fisheries in the 1950s and 1960s illustrates how subsidies can increase capacity and lead to depletion. Between 1954 and 1968, Canadian subsidies increased the capacity of the Northwest Atlantic offshore fishing fleet by more than 18 times, creating twice as much capacity as could capture fish if resources were to be used sustainably. In 1970, the Canadian government "acknowledged explicitly that its subsidies for vessel construction and modernization over the previous two decades had led to the rapid expansion of larger vessels, creating serious overcapitalisation." An Organization for Economic Cooperation and Development (OCED) study reported that subsidies also contributed to overcapacity in the fishing sector in many member states: New Zealand in the 1970s and early 1980s, Spain's Galician fisheries in the 1980s, Norway in the 1960s, the EU during the 1980s, and the United States (US) in the late 1970s and 1980s. These examples of the history of fishing fleet subsidization in industrialized countries and the resulting depletion of their stocks illustrates how overcapacity in a sector can encourage overuse of a fragile resource and lead to its exhaustion, posing threats to the surrounding ecosystem and human environment.

In addition to threatening the resources on which humans depend for their incomes, fishing subsidies may pose direct threats to crucial aspects of human and national development such as livelihood and food security. These threats are seen most strikingly in the interaction between developing nations and the industrialized countries with whom they sign access agreements. Frequent realities of EU-West African access agreements, for example, include a disregard of African fishing regulations and international agreements by distant water fleets, an inability by African states to monitor activities or catch levels of EU fleets, a lack of direct linkage between access payments and money spent on fisheries-related activities, a lack of enforcement of management efforts by African governments due to a fear of losing compensations, and the challenge of depletion of valuable fisheries stocks. There are also often many adverse effects on African coastal communities such as threats to the livelihoods of artisanal fishers and food security of those who depend of local fish stocks for protein intake. While EU payments often constitute important sources of revenue for West African governments, the security of fishing resources, vital to so many aspects of these countries, are often threatened.

The case of one West African country that signed an access agreement with the EU, Senegal, illustrates how subsidies and trade agreements can lead to threats on food and livelihood security. The Senegalese government initially implemented subsidies as a way to encourage fish production to supply domestic consumers with a protein source; fish accounts for 75 per cent of the population's protein needs. But the use of subsidies has since become incompatible with this goal since the end result has been a re-orientation towards encouraging the exportation of fisheries resources and agreements with the EU for access to Senegalese fisheries.

The combination of domestic export subsidies, which lower the costs of production for those fishers who chose to export, and trade advantages granted under the Lomé Agreement of 1982, which authorized customs exemptions to products originating in African, Caribbean, and Pacific (ACP) countries and increased the competitiveness of Senegalese piscatorial products in the European market, have encouraged small-scale fishers in the country to both intensify their efforts in response to international demand as well as orient themselves towards the export market.

These developments in subsidy policy, along with several fisheries agreements that have been signed with Japan and the EU, have threatened livelihood and food security in Senegal.

The impact of growth in subsidy-driven exports has lead to a "breakdown in the supply of cheap protein to the population," since local producers began to switch their activities towards the capture of species of high market value in European economies rather than those affordable to local consumers. The increasing scarcity of fisheries resources has also affected the traditional, artisanal fishers, whose techniques, which require low technology and employ many people, are beneficial for local resource and employment security. Since fish have become scarcer, foreign fishing fleets have become a larger presence, and local fishers lack the equipment and fuel to travel farther out to sea where the remaining catches are located, they have been forced to make deals with European or Asian boats to remain employed.

In return, locals allow the foreign boats access to areas originally reserved for their needs and use. This trend threatens local control of the resource as well as livelihood security. In a country where fish provides a vital source of sustenance for the local population, in terms of both food and employment, the way in which subsidies encourage a reorientation of resources to foreign markets is a matter of much concern for the sustainable development of the country.

CAPABILITIES OF DEVELOPING COUNTRIES IN FISHERIES

Clearly, the needs and capabilities of developing countries in fisheries vary widely. In this report we have identified the main strategic issues and trends in fisheries and aquaculture from a global perspective. The strategy that we advocate for USAID is to identify and address the common needs within and among countries to benefit the greatest number of people for the resources available.

Considering the issues and future trends we have identified in this study, we recommend the following strategic approaches to USAID:

USAID needs to substantially increase its programmatic emphasis and enlarge its *financial and human resource commitments to global fisheries and aquaculture.*

Given the importance of fish and fisheries to the global economy and their importance in poverty alleviation and food security, USAID has ample justification to increase its global profile in fisheries and aquaculture. We strongly recommend that USAID substantially increase its programmatic emphasis, funding, and enlarge its staff commitments to global fisheries and aquaculture issues.

If USAID cannot increase its core commitments in these areas, the Agency should consider augmenting its core staff by rotating into USAID mid-and senior level professionals from state and federal governments, academia, NGOs, and industry. USAID needs to make more prominent the importance it gives to the sustainability of fisheries and aquaculture at all levels of its bureaucracy: at its DC headquarters, and in all of its missions, and its regions. In addition, USAID needs to give a higher profile to capture fisheries and aquaculture issues in descriptions of its overall agriculture and natural resources portfolios. We find the word "fisheries" missing from these; and we believe aquaculture is too "buried" in lists describing a wide range of agricultural issues that interest the USAID.

USAID needs to play a central role in mobilizing America's considerable human *and institutional resources in fisheries and aquaculture to assist developing countries.*

Our vision is one in which the USAID is a world leader in channeling high quality, "needs directed" kinds of technical assistance in fisheries and aquaculture to developing countries, mainly in the form of capacity building though education and training opportunities, but also in applied research. In order to achieve this vision, we believe that the USA needs to increase its competence and competitive position in dealing with fisheries and aquaculture issues in developing countries.

USAID needs to play a central role in mobilizing America's considerable human and institutional resources in fisheries and aquaculture to assist developing countries. This can be brought about by a variety of initiatives, but the basic requirement is for US applied science and outreach professionals to work more often and for longer durations in developing country fisheries and aquaculture situations in a way that focuses more closely on collaborative solutions to common issues.

A key element of USAID approaches must be the leveraging its organizational and leadership strengths in fisheries and aquaculture by utilizing the considerable array of expertise and talent that exists in America's government (USDA, NOAA-OAR, NOAA-Sea Grant, NOAA-Fisheries, USGS, etc.), universities, state agencies, industry, and NGOs. We encourage USAID to develop innovative ideas/proposals to link with the interests of the US State Department and the Peace Corps to develop additional policy, research, and extension capabilities. In addition, the USAID needs to better insure that its multi-lateral and bilateral investments make the most use of American expertise in fisheries and aquaculture, especially in regards to instruments and relationships with The World Fish Center, various UN organizations, The World Bank, and The Asian, African, and Interamerican Development Banks.

The Pond Dynamics/Aquaculture CRSP (PD/A CRSP) is one notable example of how the USAID can play an enhanced role in mobilizing America's considerable human and institutional resources in fisheries and aquaculture to assist developing countries. The PD/A CRSP has been involved with over 50 institutions and NGOs in 27 host countries. Since the inception of the PD/A CRSP over 200 researchers have been involved and over 400 graduate and undergraduate students supported.

During the past eight years the PD/A CRSP has worked (or is currently working) with 18 US universities in 16 states. This vital capacity-building has enhanced and strengthened host countries' abilities to further develop aquaculture and provide an additional and much needed protein source to the local and regional populations.

However, USAID funds have been far too limited to develop long-term University centers of excellence in capture fisheries and aquaculture; plus the CRSP lacks a broader mandate in order to engage fully in the urgent issues of nearshore and inland fisheries, marine aquaculture, coastal area management, and the comprehensive, systems-based natural resource management approaches we recommend that would ensure a sustainable future for capture fisheries and aquaculture.

The USAID needs to examine how the CRSP, the USAID Cooperative Agreements, and other current/past administrative initiatives in coastal management and aquaculture could be used as models to expand long-term engagement in critical regional issues that have place-based centers and can involve an expanded collaboration between USAID/Government/Universities that could build on America's strengths and competitive position.

Foreign nationals need specialized short and long-term training in the US. Trained human resources are essential to resource management, which require, inter alia: multi-disciplinary expertise in fishery resource assessment; bio-economic and socio-economic analysis; management techniques; fishing technology, marketing and quality control; resource monitoring; fishery surveillance; and fisheries legislation.

The USAID Cooperative Agreements in aquaculture, fisheries and comprehensive coastal area management with American universities were very effective programmes for capacity building worldwide. These Agreements created an impressive cadre of globally important leaders—and good will—in fisheries and aquaculture throughout the academic and governmental institutions of many nations. We urge USAID to enhance its commitments to the building of additional leadership capacities in developing nations by elevating overall human resource capabilities to better manage fisheries and aquaculture. Additional Cooperative Agreements for training on a variety of concepts and skills are needed in order to have a chance at sustaining and rehabilitating fisheries and aquaculture ecosystems in many nations, to strengthen institutions, and to improve individual performance. We recommend USAID to:

- Increase funding to and the participation of American scientists in The World Fish Center, and more generally, in the current and planned activities in fisheries and aquaculture of the Consultative Group in International Agriculture Research (CIGAR),
- Participate actively in the UN/FAO Associate Professional Officer programme, so that young professionals can gain varied and broad global experiences in international settings,
- Increase the recruitment of fishery and aquaculture graduates into USAID missions and train them more extensively in pre-service at American universities,
- Fund or facilitate additional foreign student degree and certificate programmes in fisheries and aquaculture and associated resource sciences at US institutions (both at universities and at US government organizations),
- Fund or facilitate additional targeted technical assistance missions (social science, economics, fisheries management, processing, labeling, GIS, HACCP, etc.),

We recommend that USAID lead a planning process that could result in establishment of formal collaborations with a suite of US government/University centers of excellence— possibly using an expanded and better funded CRSP mechanism—in order to develop additional applied science and extension/ outreach capabilities in capture fisheries and aquaculture, and to better organize long-term, strategic and medium-term implementation plans and regular impact assessments of an expanded USAID portfolio in capture fisheries and aquaculture.

USAID needs to bridge the "digital divide" to develop solutions to fisheries and aquaculture issues in developing countries.

The Internet, supported via cable and satellite, can be an outreach pipeline, not only for spatial analyses, but also for moving more general information on fisheries and aquaculture to developing countries. There is a vast storehouse of fisheries and aquaculture information in the form of on-line technical reports and publications from US government and state agencies, NGOs, universities, professional organizations, and industry that could be "piped" abroad via the Internet. In many cases, simply raising awareness of the availability and location of the material would suffice as a useful intervention. Distance learning via the Internet, TV and radio is increasingly being used for training in developing countries. Even though, the US possesses the means to be effective in these media, training on fisheries and aquaculture using these media is not yet common. Solutions to the problems of fisheries and aquaculture share a fundamental need of basic comparative information on causes and pathways. Nearly all of the issues are not isolated; rather they are shared by neighbouring communities and countries, or topically right around the globe. Furthermore, many of them already have been experienced and treated in many areas of the world, prominently in the USA.

A fundamental problem is that the applied science that has been employed and experience gained in surmounting the issues is not yet readily available to the developing world at acceptable costs.

Therefore, an important advancement of USAID could be an initiative to help bridge the "digital divide" by increasing its emphasis on identifying, compiling and packaging solutions to fisheries and aquaculture issues of developing countries, and by broadening USAID involvement with many organizations specialized in technical, social and economic solutions in fisheries and aquaculture for which the USA has a comparative advantage—especially by taking advantage of the latitude within USAID for developing a broad variety of initiatives that include the US federal government and state agencies, commercial firms, NGO's, Land/Sea Grant universities, regional aquaculture centers, and international organizations. Several kinds of USAID initiatives are required to:

- Synthesize, package, and deliver applied research information by enhancing extension systems to move information to developing countries and to disseminate it in an effective manner;
- Assess the potential of distance learning to significantly improve technical and managerial competences in fisheries and aquaculture in developing countries;
- Facilitate information flow via choosing methods of information delivery that are appropriate for different audiences in different regions (increasing Internet access, radio, CDs, DVDs, videos, etc.);
- Sponsor collaborative in-country research between US professionals and applied fisheries and aquaculture professionals in developing nations on problems that are indigenous to the country or region.

DEVELOPMENT OF THE AMAZON AQUAFORUM

We applaud the Pond Dynamics/Aquaculture CRSP in its development of the Amazon Aquaforum, an Internet-based information exchange aquaculture network in South America which supported development of a variety of technologically appropriate information delivery methods. The CRSP also maintains an invaluable, comprehensive, standardized database of information collected from throughout the world. The CRSP developed a web-based resource for small and medium-scale farmers in Latin America called the Web-based Information Delivery System for Tilapia (WIDeST), a decision-making tool that enables users to gain access to useful resources when deciding on appropriate, site specific methods and aquaculture practices.

USAID should prioritize the improved management of coastal marine and inland fisheries by providing technical assistance to evolve innovative fisheries management schemes in developing countries including but not limited to: property rights, co-management, and the use of marine protected areas; plus assist in the development of more accurate and reliable fisheries and aquaculture data reporting systems.

USAID needs to invest additional resources to assist in achieving the long-term goal of sustaining the world's invaluable marine and inland capture fisheries, which are disproportionately located in developing countries. Additional investments in capture fisheries could positively impact and better leverage USAID's current and planned investments in coastal area management and aquaculture.

Additional assistance is needed to engage users and government institutions to bolster management and governance structures to address issues of overcapacity, access, marine tenure, critical habitats and nurseries, and a priority range of social issues such as gender relations. Investigations are needed into conservation engineering (innovative gears and management); roles of reserves, protected and conservation areas; plus investigations regarding how best to protect freshwater flows to estuaries (and rivers to lakes)—especially the timing, volume, quality and pulsing of freshwater flows to critical estuaries.

In addition, the USAID could play a major role in analyzing and promoting effective fisheries management, government policies, reforms, market/trade policies, and reductions of subsidies. In comparison to the long-standing emphasis on technical investments in fisheries for stock assessments, etc., comparably little attention has been paid to the users of fishery resources. This is despite the fact that the observed successes, failures, and constraints experienced in marine and inland capture fisheries management are social and economic in nature.

Another problem is that fishery researchers often have a low status and income in a given national context, have limited facilities and resources, few opportunities for in-service training, and have limited access to outside scientific research information. We recommend that the USAID expand its investments in the community-based management of marine protected areas, property rights, and overall investments in social ecology and ecosystems-based management of fisheries.

Illegal, unreported and unregulated fishing is found in all capture fisheries, irrespective of the location, species targeted, fishing gears employed or level and intensity of exploitation. USAID should work with the reporting countries to improve fishery statistics, primarily to meet national needs with regard to food security and fisheries management. Unlike capture fisheries, the separate monitoring of aquaculture is relatively new in most countries, and often there are less well-established systems of data collection as compared to capture fisheries.

USAID needs to substantially increase its support to develop more comprehensive, sustainable, ecologically and socially compatible, environmentally-friendly and economically viable aquaculture systems in developing countries that have the long-term goals of poverty alleviation and food security.

To meet global demands for fisheries products, aquaculture will continue to grow at a rapid pace over the next 10 years; then, its rate of growth will slow until the considerable environmental constraints it faces are solved. Aquaculture faces a number of important problems, including access to appropriate technologies, lack of comprehensive, inter-sectoral planning, and a lack of financial resources for the poor, information on its environmental and social impacts and diseases. The priority areas for further applied research support include:

- Land, water and feed/nutrient use in aquaculture in comparison with other animal protein production systems;
- Sustainable intensification and non-consumptive water use in freshwater aquaculture production;
- Participatory management approaches to the comprehensive development of aquaculture ecosystems as sustainable means of rural development;
- Sustainable coastal aquaculture development, especially technologies that avoid user conflicts;
- Social and economic research to add insights into the adoption of aquaculture by poor rural households;
- Genetically advanced technologies for sustainable stock enhancement and ranching programmes, plus the domestication, selective breeding, and genetic improvement of existing aquaculture species;
- Technologies to solve disease problems and innovative management solutions to improve the health of aquatic animals;
- Development of low cost, non-fish meal based feeds;
- Training in the quality and safety of aquaculture products; and
- Research in making emerging technologies cost-effective, including recirculating systems, and offshore aquaculture systems.

The appropriate role for the USAID in fisheries and aquaculture biotechnology is to support applied research and outreach activities that engage in well-known, conventional genetic improvement techniques, such as selective breeding, etc., as opposed to research support to advanced biotechnologies, such as transgenics. USAID support for The World Fish Center's programme on the genetic improvement of farmed tilapias (GIFT)—and the development of international protocols for product dissemination—is an excellent example of the types of biotechnology investments the merit USAID's future consideration. However, it is also very important that USAID engage nations who have large and active programmes in advanced fisheries/aquaculture biotechnology—such as China—in issues of policies, protocols, environmental, market and other social impacts of transgenics.

USAID should prioritize its assistance to capture fisheries and aquaculture activities that are more integrated, comprehensive, community-based, and use "systems approaches"—such as ecological and integrated farming/fishing systems research and extension approaches—in both rural and urban settings. The current agriculture emphasis of USAID is on plant commodity research, not on a comprehensive, agro/aqua-ecosystems research/extension approach. We urge the USAID to support long-term, applied research and development that makes expanded use of participatory ecological and social science tools to empower community control of fisheries and aquaculture systems; and to better integrate aquaculture and fisheries activities into the comprehensive management of natural and social resources of its missions, target nations and regions.

The kinds of assistance provided by USAID should be based on a combination of: (1) assessed needs and capabilities in developing countries, and (2) the comparative advantages held by the USA in technical expertise, education, communications, business management and commercial products. Assistance focused on US comparative advantages provides a way to get around duplication of effort among competing international organizations, while still fully supporting the Code of Conduct for Responsible Fisheries. The top priority for USAID should be the sustainable development of community-based, integrated farming and fishing systems with the long-term goals of poverty alleviation and food security.

AQUACULTURE AND FISHERIES ACTIVITIES

Aquaculture and fisheries activities are frequently poorly planned and considered; funded separately from activities that greatly impact them; and are generally neglected by all levels of government. Fisheries activities should be planned as a continuum, and comprehensively as part of the planning for integrated natural resources management (water, wastes, agriculture, etc.), and the sustainable development of human communities.

Fisheries and aquaculture information should be linked/coordinated with activities in other sectors (*e.g.*, agriculture, forestry, food processing, distribution, etc.). Fisheries and aquaculture should be planned up-front—not as an afterthought—in all water resource development projects such as reservoir and irrigation projects. Coordination is both justified and essential because of shared issues with other sectors of sustaining biodiversity, maintaining water quality and quantity, the management and development in coastal areas and in river and lake basins, and addressing environmental degradation, mitigation and restoration, and the need to improve governance. USAID is encouraged to prioritize its support to applied research and development activities that articulate well with the natural and social resource and farming systems contexts of a nation/region, and (a) plan for the activities as one part of a comprehensive management strategy for the non-consumptive, multiple uses of water; and (b) use ecosystems-based management approaches that promote the more comprehensive, long-term stewardship of marine and freshwater environments. We urge the USAID to support the expanded use of participatory tools to empower community controls over fisheries and aquaculture systems; to use innovative co-management methods to sustain local water and coastal resources, and the community-based management of water bodies for fisheries and aquaculture. Fisheries and aquaculture provide foods of very high nutritional value for households.

When resource-poor fishers and farmers combine fisheries, agriculture, aquaculture and the conservation/rehabilitation of natural resources in innovative "ecosystems approaches"' they improve their food supplies, increase their incomes, and become better able to withstand environmental and economic fluctuations; thereby decreasing risks, increasing fishing and farm sustainability, and contributing greatly to rural social and economic development.

We encourage the USAID to incorporate additional, applied social science and micro-economics research into innovative ecosystem-based methods that empower communities to better manage and control fisheries systems, and to develop more environmentally and socially compatible aquaculture systems (protected areas, property rights, innovative co-management approaches, etc.). Most of the modern aquatic resource crises have roots in social issues that are poorly known, such as the "shifting nature of modern survival" in developing nations. Millions of people do not only fish or farm, rather, they derive income from multiple activities and sources, and in some cases, conduct long distance seasonal migrations between inland farming systems and fisheries systems, and vice versa. USAID is encouraged to support the development of aquaculture systems that are well integrated into existing water resource systems, are virtually non-consumptive of water, and make multiple uses of water.

SMALL BUSINESS DEVELOPMENT

Aquaculture has the potential for small business development but additional technical assistance is needed by business professionals familiar with economics, labour dynamics, opportunity costs, issues of price and volume competition, and other commercial and competitive contexts of other sectors such as agriculture, etc.

The recent activities of the Pond Dynamics/Aquaculture CRSP (PD/A CRSP) are notable in this regard. In 2002, the CRSP conducted regional meetings in Latin America, the Caribbean, Africa, and Asia, plus commissioned a report to explore the current status of aquaculture in Eastern Europe/Central Asia. Participants represented diverse areas of expertise; gender diversity was also a criterion of panel composition.

During the meetings, participants were asked to identify and prioritize constraints to aquaculture development in the region of their expertise. Three central needs-directed programme areas emerged after analyzing the results of the meetings. This movement of the PD/A CRSP to become more of a "system-oriented" network as opposed to a "commodity" collaborative is noteworthy and laudable; and if additional resources were available, these concepts could be developed further.

We cannot emphasize more strongly that poorly-funded, short-term projects with broad, "global" goals will not make lasting impacts on the conservation and sustainable development of fisheries and aquaculture systems in developing countries; and that these approaches do not serve the strategic interests of the United States. Research, education and extension assistance must target the long-term engagement of institutions, and deliver approaches, findings and insights towards these institutional systems and organizations that will continue the long-term engagement with farmers/fishers that can weather the invariable fluctuations in development assistance that will occur over time.

USAID needs to develop comprehensive strategic and implementation plans and regular impact assessments of an expanded capture fisheries and aquaculture portfolio. USAID missions and regions worldwide should include capture fisheries and aquaculture into their strategic plans for the management of natural resources—or they will be incomplete—especially in regards to USAID plans for involvement in the issues of water allocation and quality, and plans for the comprehensive management of marine and inland coastal areas.

USAID needs to sustain a strategic, long-term commitment to fisheries and aquaculture as parts of comprehensive natural, aquatic resource management. Short-term projects should be part of larger, longer-term strategic frameworks for directed action that have adequate accountability to measure strategic progress and impacts. USAID needs to conduct regular, transparent processes that result in the publication of strategic and implementation plans for fisheries and aquaculture over 5 to 10-year time frames. The movement from short term projects to longer term investments in centers of excellence, for example, will require the USAID to develop strategic planning and assessment processes that are much more comprehensive and "living". With more stable investments strategic plans have a shorter "shelf life" and require more frequent review and evaluation in order to make the necessary "mid-course changes". We recommend a process similar to that used by Standing Committees of the U.S. National Research Council where external expert advisors regularly measure progress on investment portfolios.

Within the fisheries sector, USAID should take advantage of shared needs for land and water resources and issues in common to identify sub-sector groupings for directed technical assistance and development activities. We identify two groups that require distinct interventions: (1) offshore (*i.e.*, within the EEZs) and high seas fisheries share common issues of resource and fisheries management that include shared resources, over harvesting, and excess capacity, and (2) coastal and inland fisheries as well as aquaculture to a great extent, may interact both positively and negatively with one another, sometimes competing for space, resources and markets, but they also share many common issues that are external to the fisheries sector, most importantly the environment and poor governance.

PRINCIPAL DIRECTIONS OF BIOTECHNOLOGICAL DEVELOPMENT

In a recent review of agricultural technology, the Office of Technology Assessment (OTA) of the US Congress defined biotechnology to include 'any technique that uses living organisms or processes to make or modify products, to improve plants or animals or to develop micro-organisms for specific uses-it focuses upon recombinant DNA and cell fusion technologies'. Longworth concurs with this definition, with the significant addition of tissue-culture techniques, an aspect of biotechnology which has great economic, and therefore social and political, potential impact. Despite debates about whether these or any other definitions are adequate, they do touch on the main aspects of biotechnology which deserve to be considered.

The aspect of this new biotechnology which most captures the imagination and stirs the greatest controversy is gene splicing or recombinant DNA techniques, which inspire researchers to consider the possibilities of producing reproducible animals and plants markedly different from current existing species, and referred to as transgenic species. Already in higher animals and plants recombinant DNA has produced transgenic forms which are being commercially exploited. Their agricultural significance is, however, so far limited, and commercial application is concentrated in highly profitable pharmaceutical and horticultural markets. Genes have been introduced into several animal species which alter their protein synthesis to enable transgenic sheep to produce insulin in their milk, and rabbits to produce interferon.

A completely different application has originated in Denmark for salmon, where it has proved possible to introduce germplasm which enables the salmon's physiology to handle heavy metals, which are normally toxic, so opening up new locations for farm fisheries. Much current research is directed to conferring disease immunity on animals, and holds out the prospect of widespread commercial application. In plants one achievement has been the transfer of genetic resistance to antibiotics in the petunia, and another has been the introduction of storage-protein genes from French bean plants into tobacco plants.

As yet commercial progress with recombinant-DNA technology in plants appears limited, but extensive opportunities beckon, particularly because plant research is less restricted by the ethical and animal-welfare concerns which apply to research on transgenic animals. The most important commercial developments based on gene splicing have so far occurred with genetically much simpler microorganisms, and it is with these that the greatest short-to mediumrun commercial potential lies. Already genetically engineered micro-organisms are producing a variety of hormones, vaccines, enzymes and other proteins. Important examples are the production of insulin, vaccines for neo-natal diarrhoea in calves and piglets, and the bovine growth hormone BST identical to that produced in cows which can stimulate a 20 per cent increase in milk yield. BST is already a source of problems for legislators in the European Community, and has provoked strong reactions from the media and milk consumers.

As far as larger animals, and cattle in particular, are concerned, it is developments in embryo transfer and in many processes for manipulating reproduction which hold out the prospect of continuing increases in yields, feed conversion efficiency and general economic efficiency. Already, apparently over 1 per cent of dairy calves in the USA are from embryo transplants, despite the high costs still associated with this procedure. Widespread adoption of these sophisticated technologies would further distance livestock farming from its traditional rural simplicity, and from the natural mating of animals to produce offspring. It would place technical demands upon the operators which favour large-scale company farms capable of supporting a range of highly trained specialists.

Longworth identifies new techniques of tissue and cell culture as having 'the potential for enormous advances in crop improvement in the next couple of decades'. Those techniques 'can both increase the genetic diversity and greatly increase selection efficiency', and they permit innumerable plants to be reproduced asexually from single cells or small pieces of tissue. The particular technique known as 'callus culture' has been used for many years to clone highly valued horticultural plants such as orchids, and cloning of cuttings is widely practised for tree crops such as tea and palm oil as well as by millions of gardeners for garden plants.

According to Longworth 'cell culture' of single cells has unexpectedly, and so far inexplicably, resulted in plants with different properties being regenerated from the same clump of parent tissue. Sugar cane, maize and potato plants regenerated in this way have been found which are resistant to important pathogens. This application of cell culture with the capacity to generate vast numbers of seedlings rapidly has the potential to simplify greatly the hitherto labourious procedures of plant breeding and selection.

However, in terms of current and immediate commercial importance, it is through micro-organisms that biotechnology has its greatest impact. Microbial fermentation processes have been of great commercial significance for centuries, for example in bread, wine and cheese production, and there is the prospect of considerable development. Two recent examples of new processes indicate the sorts of impact that such developments can have. In the late 1960s genetically engineered bacteria were developed which were able to digest corn starch to produce high-fructose corn syrup (HFCS) and which left as a residue corn gluten, which is now an important protein feed for livestock. HFCS has made considerable inroads in the United States and in Japan. It has largely replaced sugar as a sweetener in Coca Cola as well as many other food and drink products.

This has helped depress sugar prices to Third World growers. In the European Community steps have been taken to prevent HFCS and other new powerful sweeteners from being produced and from undermining the market for domestically grown beet sugar, which is supported by the Common Agricultural Policy. A second important application of microbial fermentation has been the production of ethanol from sugar cane in Brazil and from maize in the USA. The commercial viability of these processes is critically dependent upon the price of oil as the main non-renewable source of fuel, and, to date, massive subsidies have been required to maintain the Brazilian and USA ethanol programmes.

Longworth states, however, that a new biotechnology, Sucrotech, is being patented, which will not only reduce the cost of producing ethanol from sugar cane, but will simultaneously produce fructose at a cost which will be competitive with HFCS and may thereby reclaim part of the sweetener market for sugar cane. That such possibilities are in prospect indicates how volatile the future might be; biotechnology has tilted the competitive balance from sugar cane to maize, causing economic pressure and even disruption to sugar-cane-dependent economies and may in future switch it back again.

In the long run the capacity to ferment fuels microbially from renewable agricultural feedstocks points to an important long-term reorientation of agriculture if non-renewable oil becomes uncompetitively expensive as a fuel for cars and feedstock for certain chemicals. It suggests that eventually there will be an increased emphasis on agricultural production of industrial feedstocks at the same time as continually increasing food output will be required to feed the expanding world population. All this will require considerable increases in agricultural productivity, to which biotechnology will increasingly contribute, but it will at the same time impose great strains on the natural environment and the structure of agriculture.

PRIVATE-SECTOR CONTROL OF BIOTECHNOLOGY DEVELOPMENT

It has been the role of the public sector to undertake research and development of agricultural technology as a public service to firms which might profit from translating that R & D into a commercial product or process, to farmers who might profit from adoption of the technology, and perhaps most importantly to consumers at home and abroad who benefited from the lower prices resulting from greater abundance. This was particularly true of the phase of change dominated by improvements in plant and animal breeding, the 'old' biotechnology. While the public sector had a role in basic research for chemical and mechanical technologies, the benefits of research expenditure in these areas were easier to capture by private companies investing in R & D, so that increasingly the public sector has taken a smaller role in these areas although maintaining a strong regulatory role in regard to agricultural chemicals in particular.

Where it is impossible to prevent others from escaping payment for the research costs, either because the product is easily copied (seeds which can be regenerated by farmers or other firms) or because proposals for new methods can be readily implemented, private firms are understandably unwilling to invest. Machines, insecticides, fungicides and other manufactured inputs do lend themselves more readily to private exploitation.

Nevertheless the returns to R & D in these products do depend upon the degrce of difficulty potential competitors would have in copying the product. In some cases there are inherent technical difficulties in copying thc process, or it would be prohibitively expensive, but in other cases it is the ability to obtain patent protection which creates legal barriers to potential competitors' ability to become 'free-riders' and which protects incentives to private investment in R & D.

Traditionally, however, it has been impossible for plant and animal breeders to obtain patent rights for their products, which is one reason why public-sector R & D has remained so important in this area. For, as under the European Patent Convention (EPC) of 1973, it has been judged impossible for plant varieties to satisfy one of the key criteria to qualify for a patent, namely proof of 'an inventive step'. The application of standard breeding practices to generate new varieties by crossing existing plants or animal strains has not been deemed to be invention. Thus under the EPC one set of exclusions from patentability is 'Plant or animal varieties or essentially biological processes for microbiological processes or the products thereof.'

In the absence of patent rights plant-breeding firms in particular have worked hard to obtain other means of protection and royalties for their products. The history of the development of Plant Breeders' Rights (PER) is presented by Mooney, and is of particular interest because of the concerns he expresses about the consequences of allowing the basic genetic stock, which underpins agriculture and hence our whole society, from becoming private rather than public property. While Mooney's concerns are expressed in relation to 'old' biotechnology varieties of plants, they are of particular importance with respect to the products of the 'new' biotechnology. For bio-engineered products are capable of meeting the inventive-step criterion of patentability, and both micro-organisms and transgenic animals have now been patented in the USA.

Before examining developments in the seed industry it is worth touching upon the implications of changes in research policy which stress increasing reliance upon private R & D to develop and exploit the new biotechnology and other technologies.

In the UK government has decided that it should withdraw from funding what it terms 'near-market research', that is, research beyond the basic phase and which is preparation for commercial exploitation. This policy is based in part upon the arguments that industry should be investing more heavily in R & D, and that research strategies at the near-market stage should be driven by assessments of likely commercial success which can best be made by the firms involved, and will therefore lead to greater efficiency in allocating research funds.

This has led to the closure of a number of government-financed agricultural research institutes, the scaling down of others and the sale of the National Seeds Organization by auction to Unilever; Unilever won against competition with BP and other major public companies.

This deliberate attempt to switch an increasing proportion of agricultural R & D expenditure from the public to the private sector carries with it a number of risks. In the first place there is a controversy about whether there is under-investment in R & D so that the returns to extra expenditure are high, or whether the converse is the case.

If there is under-investment, then withdrawal of public support for applied research will exacerbate this, as the private sector will only undertake R & D expenditure on those products and processes from which exclusive benefits can be captured by the investor.

Thus in relation to biotechnology in the USA, where again public support for applied biotechnology research is relatively small, Stallman and Schmid (1987) argue that there will be emphasis on technologies which are applied in factory conditions where secrecy and control can be maintained. For technologies which cannot be confined to factories, such as seeds, Stallman and Schmid state that 'Firms are also considering mechanisms which "scramble" the genome of a plant in the second generation', in order to prevent farmers from reproducing seed with the enhanced, engineered characteristics. Clearly such actions are designed to frustrate the maximum spread of benefits from the technology and to maximize private profit for companies investing in research.

Another facet of commercially orientated biotechnology research is that it will aim at the most important crops and livestock products, those produced by the largest farming units, and those which are most heavily subsidized. In the latter case this will worsen the budgetary problems of adjusting agricultural policy in OECD countries. The probable neglect of minor crops, difficult habitats and small farmers means that the public sector will have a defined role in agricultural technology research, but one in which it will be relegated to the second division and where it is unlikely to prove successful in terms of the commercial yardstick of rates of return which is increasingly emphasized by public-research policy.

TECHNOLOGICAL PROCESS IN GENETIC ENGINEERING

In order to understand how genetic manipulation is accomplished, it is important first to understand the structure of deoxyribonucleic acid, or DNA. Within its chemical structure, DNA stores the information that determines an organism's hereditary or genetic properties. DNA is made up of a linked series of units called nucleotides (Blaese), Different nucleotide sequences determine different genes genetic information. Genetic engineering is based on this genetic information.In order to understand how genetic manipulation is accomplished, it is important first to understand the structure of deoxyribonucleic acid, or DNA. Within its chemical structure, DNA stores the information that determines an organism's hereditary or genetic properties. DNA is made up of a linked series of units called nucleotides (Blaese), Different nucleotide sequences determine different genes genetic information. Genetic engineering is based on this genetic information.

Genetic manipulation is carried out through a process known as recombinant-DNA formation, or gene splicing. This procedure behind genetic engineering is one whereby segments of genetic material from one organism are transferred to another. The basis of the technique lies in the use of restriction enzymes that split DNA strands wherever certain desired secjuences of nucleotides, or specific genes, occur. This desired segment of DNA is referred to as donor DNA. The process of gene splicing results in a series of fragments of DNA, each of which express the same desired gene that can then combine with plasmids (Rubenstein).

Plasmids are small, circular molecules of DNA that are found in many bacteria. The bacteria act as vectors in the process of genetic engineering. The desired gene cannot be directly inserted into the recipient organism, or host, therefore there must be an organism that can carry the donor DNA into the host. Plasmid DNA is isolated from bacteria and its circular structure is broken by restriction enzymes (Dworkin). The desired donor DNA is then inserted in the plasmid, and the circle is resealed by ligases, which are enzymes that repair breaks in DNA strands. This reconstructed plasmid, which contains an extra gene, can be replaced in the bacteria, where it is cloned, or duplicated, in large numbers. The combined vector and donor DNA fragment constitute the recombinant-DNA molecule. Once inside a host cell, this molecule is replicated along with the host's DNA during cell division. These divisions produce a clone of identical cells, each having a copy of the recombinant-DNA molecule and thus permanently changing the genetic makeup of the host organism (Steinbrecher). Genetic engineering has been accomplished.Genetic manipulation is carried out through a process known as recombinant-DNA formation, or gene splicing. This procedure behind genetic engineering is one whereby segments of genetic material from one organism are transferred to another. The basis of the technique lies in the use of restriction enzymes that split DNA strands wherever certain desired secjuences of nucleotides, or specific genes, occur. This desired segment of DNA is referred to as donor DNA. The process of gene splicing results in a series of fragments of DNA, each of which express the same desired gene that can then combine with plasmids (Rubenstein).

Plasmids are small, circular molecules of DNA that are found in many bacteria. The bacteria act as vectors in the process of genetic engineering. The desired gene cannot be directly inserted into the recipient organism, or host, therefore there must be an organism that can carry the donor DNA into the host. Plasmid DNA is isolated from bacteria and its circular structure is broken by restriction enzymes (Dworkin). The desired donor DNA is then inserted in the plasmid, and the circle is resealed by ligases, which are enzymes that repair breaks in DNA strands. This reconstructed plasmid, which contains an extra gene, can be replaced in the bacteria, where it is cloned, or duplicated, in large numbers. The combined vector and donor DNA fragment constitute the recombinant-DNA molecule. Once inside a host cell, this molecule is replicated along with the host's DNA during cell division. These divisions produce a clone of identical cells, each having a copy of the recombinant-DNA molecule and thus permanently changing the genetic makeup of the host organism (Steinbrecher). Genetic engineering has been accomplished.

Genetic engineering is a technological process for undertaking activities which do not, and cannot, occur in nature. In this sense it is perfectly legitimate, therefore, to label genetic engineering as ‘unnatural’. The European Community has incorporated this concept in its definition of genetically modified organisms, and it has been adopted by the UN Environment Programm (in the working of its Fourth Expert Panel set up under the Biodiversity Convention as part of the follow through to the United Nations Conference on the Environment and Development held in Rio de Janeiro in June 1992): ‘organisms in which the genetic material has been altered in a way that does not occur naturally by mating and/or natural recombination’. Thus, hybrids and other modified organisms obtained by traditional breeding techniques are excluded from the definition of genetically modified organisms.

It is important to note that the process as well as the products are novel. The promoters of the new technology believe that a concern for the process itself is specious, downplaying its novelty whenever confronted by discussions of regulatory oversight. However, one of the bases of the decision of the US Supreme Court in the *Chakrabarti* case (which, 5 to 4, upheld the patentability of genetically engineered microbes, and which has been unquestionably accepted by foreign governments as disposative of the issue of patentability of genetically engineered life forms) was that the element of 'novelty' required by the patent law was to be found in the process by which the microbe had been created.

One reason why we cannot ignore the powerful novel aspects of the processes of genetic manipulation is because we are not omniscient as to what will, in fact, happen when we alter genomes. Genetic manipulation is not like a child's game of Legos or Tinkertoy, in which parts can be rearranged or linked up with only simple mechanical and relatively predictable consequences.

In calculating any risk from a transgenic organism, one should consider four elements: the host organism, the foreign genes, the interaction between the foreign genes and the rest of the genome, and the environment in which the organisms will be used.... [in regard to the last two elements] the literature contains many examples of genetic manipulations where inserted genes did not respond in their new environments the way they did in their old ones or where alterations with one part of the genome caused surprising activity in other parts of the genome.

There is an element of arrogance to scientific assertions which assure us that the process itself poses no risks, since scientists know so little about actual ecosystems; for example, I have been told by agronomists that over 80 per cent of the organisms which can be identified in a soil sample from my garden are completely unknown to the scientific literature. Indeed, the Ecological Society of America itself has specifically warned about the problems presented in our lack of ecological knowledge and the resulting consequences from releasing (intentionally or accidentally) genetically altered organisms into ecological systems.

FOODS

In May 1992, the US Food and Drug Administration, responding to corporate pressures to remove the prospect of regulation of genetically altered foods, issued a set of rules which have largely left responsibility for protecting the public health and safety in the hands of the industry, permitting products to be marketed without scrutiny unless the industry indicated to the agency that it believed governmental oversight was justifiable.

However, the proposal as published by the FDA in the Federal Register perversely noted several important problem areas: creating allergens, additions of genes from sources which might violate religious and cultural norms (of vegetarians, Jews or Muslims, etc.), and implications for animal welfare and well-being (for example, the incorporation of human growth hormone-producing gene into a pig's genome produced a highly arthritic animal). Indeed, the commissioner of the FDA, David A. Kessler, and his colleagues noted the possibility that genetically engineered food might 'contain high levels of unexpected, acutely toxic substances'. In order to begin to address ethical issues presented by the genetic modification of foodstuffs, we need to have some ideas of the purposes for which these modifications are performed. (For example, if we wish to approach an analysis on utilitarian cost-benefit grounds.) Despite a great number of general statements, human nutrition and hunger do not appear to be the actual driving forces behind the development of genetically engineered foods. Other than claiming increased shelf life (which can be indirectly related to nutrition in the sense of reducing spoilage or the consumption of foods that have begun to spoil), there is hardly any indication at all that genetic engineering is being directed to create nutritious substances out of non-nutritious ones; similarly, there are few real instances of genetic engineering reducing the cost of foodstuffs, or increasing the quantities of foods actually available to populations that are hungry because of the non-existence of consumables (as opposed to hungry because they lack money to buy sufficient foods, no matter how produced). Indeed, early genetic manipulations seemed perversely designed to minimize the achievement of such goals: the creation of 'herbicide tolerant plants' which do not reduce the applications of dangerous chemicals to agricultural fields (which might occur if genetic manipulations of food crops were designed to make them resistant to insects, fungi, or disease) but instead permit higher levels of chemical application, or the introduction of recombinant bovine growth hormone to produce more milk at a time when developed countries suffer from milk gluts and have instituted programmes to kill cows and physically dump milk (milk as a commodity is price inelastic; quantity increases do not lead to price decreases).

The goals of genetically engineering foods seem to be more closely tied to increasing the economic gain and power of the corporations involved in food production (for example, the development of herbicide-tolerant farm crops has been led by corporations which manufacture herbicides), making production easier for corporations, reducing or altering packaging and transportation costs for the corporations, etc.

Environment

Industrial societies around the world are faced by massive pollution legacies from the previous emphases on chemical and nuclear industrial activities. The inability of the US Department of Energy to find a suitable site for a longterm nuclear waste repository illustrates our very poor track record for dealing with the scientific, technical, economic, and socio-political aspects of prior technological 'revolutions'. The environmental problems posed by genetically engineered organisms are likely to be substantially more intractable than those posed by these earlier instances of pollution because genetic wastes multiply, migrate, and mutate. A genetically engineered organism once free in the environment is impossible to recall. (Impacts which are irreversible must be considered with higher scrutiny than those which can be undone.)

Examples of environmental problems which need to be assessed include the risks of transgenic crops themselves becoming weeds; the risk of gene flow to wild relatives which might become weeds or pests; the growth of antibiotic resistance in species (particularly animals), since resistance genes are used as markers for the genetic engineering sites; the problem of exotic or non-native species taking over ecological niches (such as gypsy moth, starling, kudzu, rabbits in Australia, etc.) leading to extraordinary economic losses as well as ecological ones; the restriction (rather than increase) of biodiversity by selective advantageous breeding of transgenic organisms as opposed to natural ones, or the predatory results of expanding exotic species (such as the Dutch Elm micro-organism destroying the American Elm and severely restricting the biodiversity of certain areas in the Northeast United States); health issues (both of plant and animal species, as well as humans — worker health and safety as well as community security); and the occurrence of the completely unexpected, the inability to eliminate uncertainty (for example, the late 1993 floods in the Mississippi valley included the flooding of a field of genetically engineered corn and the dispersal of this plant material to unknown sites within the thousands of square miles of downstream flood plain).

Prospective ecological assessment is not being performed, despite the fact that its need has long since been recognized. And economic priorities, particularly those accruing directly to the promoters, as well as the nebulous spur of 'competition' with foreign countries, oftentimes overwhelms any interest in doing environmental impact analysis.

CONCLUDING COMMENTS

The US FDA has been praised for its science-based regulatory approach to evaluating the safety of products developed using rDNA technology based on the characteristics of the food product, and not the means by which it has been created. On the other hand, the regulatory approach of both the USDA and US EPA has been criticized for discriminating against the products of rDNA technology. A consortium of 11 major international scientific societies and the Council on Agricultural Science and Technology (CAST), an organization representing 36 scientific and professional groups, have strongly criticized the US EPA for their proposed approach to regulating plant-pesticides on the basis of the process used in its development rather than the characteristics exhibited by the plant. More recently, the US Congressional Committee on Biotechnology released a report entitled 'Seeds of Opportunity' recommending revisions to the USDA and US EPA approaches to the regulation of GM plants. It was felt that the proposals by both agencies failed to take into consideration the current scientific consensus on the potential risks associated with agricultural biotechnology. Overall, the report was supportive of the science-based policy of regulation, based on the characteristics of a food product and not on the means by which it was created.

As a part of their role, each federal regulatory agency involved in assessing the human health and environmental safety of GM crops attempts to ensure that only the most current scientific data, information, and methods are referenced. The US FDA consults a Food Advisory Committee, the US EPA refers questions to Scientific Advisory Panels, and the USDA has recently established its own advisory panel. Aside from periodic reviews by credible independent research organizations such as NAS, the individual policies of each agency have been subject to regular review for harmonization with approaches recommended by international standard-setting bodies, such as OECD, WHO and FAO. It is expected that many of the future food products of rDNA technology in agriculture may not be equivalent in composition to their conventional counterparts. However, the existing decision-tree approach to regulation anticipates these types of improved food products, and provides for the case-by-case assessment of their safety. Field testing and pre-market review for food safety provide the required level of assurance that the foods derived from the application of rDNA technology in agriculture are at least as safe as existing foods, and are consistent with all existing standards of food safety.

All new varieties of vegetables, fruit, corn and soybeans developed using rDNA technology have been subject to considerable scientific and regulatory review by the US FDA, the USDA and/or the US EPA. As such, the developers of new crop varieties have had to take into consideration not only the interests of farmers, food processors and consumers, but also the expectations of regulatory authorities.

Both regulators and consumers have had a number of opportunities to influence the acceptance of a new crop variety, and continue to have significant influence beyond final regulatory approval and commercial introduction. It is expected that the issues of human health and environmental safety will continue to be of major concern in the future as a greater number of GM food products reach later stages of commercial development. A recent trend has been observed towards the commercialization of products that offer more transparent benefits and value to consumers. The new products of rDNA technology will be expected to further challenge the flexibility that has been exhibited by the US regulatory process. As a result, both industry and regulatory authorities will need to be responsive to uphold the human health and safety record achieved for GM product introductions. The vast majority of the new products of rDNA technology are not expected to be commercially available until after 2005.

The crops and foods improved through rDNA technology have been developed with much more precision, and their human health and environmental safety has been assessed in more depth and detail than any other crops and foods developed using more conventional means. They are routinely evaluated for any potential impact on both the environment and food safety. Despite years of experience and research, no verifiable data implicating rDNA technology as a food safety or environmental hazard have presented themselves.

The research and development of GM crops have been performed under extremely controlled conditions and tight regulatory oversight from research and development within the labouratory through field trials and commercial planting for food-use. Based on evaluations conducted to date, no evidence exists to suggest that GM foods currently sold on the market pose any significant human health concerns or that they are in any way less safe than foods derived from crops developed through more conventional means.

Although the foods developed using rDNA technology have been subject to rigorous scientific review for safety, the US regulatory system has allowed for considerable advancement in the application of the technology, and significant product innovation. To continue to be successful, it is recognized that the US FDA, USDA and US EPA will need to cooperate: the approval process must be clear, timely and transparent; emphasis must be on science-based/risk-based regulatory assessments and decisions; the regulatory process must be flexible to adapt to new information; and, the agencies must be staffed by full-time, highly qualified scientists in many diverse fields.

Over the past 25 years, the approach to regulating the products of agricultural biotechnology has evolved in response to the experience gained during the research and development of applications for rDNA technology. In the future, as the number of products developed using rDNA technology continues to increase, additional research will be needed to better evaluate any potential effects on human health and the environment, and to further refine the scientific basis of making regulatory decisions.

Within the United States, the regulatory approach to evaluating the safety of products developed using rDNA technology continues to evolve as new applications continue to emerge and our understanding of the potential risks improves. To date, existing regulations have proven adequate to assure the safety of all new products developed using rDNA technology, and additional innovation has not been discouraged. Accordingly, the National Research Council has recommended that any new rules for the regulation of GM plants be sufficiently flexible enough to reflect improvements in scientific understanding. In the future, as rDNA technology finds broader application, and the new prospects for this technology become commercial realities, the regulatory approaches to evaluating the safety of GM food crops are expected to evolve accordingly.

4

Marine Fisheries and Biodiversity

CURRENT STATUS OF MARINE FISHERIES

In general, considered on a single-species basis, many marine fisheries are fully exploited or overexploited, while relatively few seem to have the potential for increased exploitation. In general, this is true both for the United States and globally, especially in estuarine, nearshore, and continental-shelf fisheries, which produce approximately 75 percent of the world's fish catches.

The primary source of global information about the condition of fisheries is the Food and Agricultural Organization (FAO) of the United Nations. In the United States the task of carrying out assessments has been primarily the responsibility of the National Marine Fisheries Service (NMFS) of the National Oceanic and Atmospheric Administration, U.S. Department of Commerce. Regional fishery management councils (established under the authority of the Magnuson-Stevens Fishery Conservation and Management Act of 1976) also are involved in the assessment of fish stocks in federal waters. Interstate fishery commissions in the Atlantic, Gulf, and Pacific regions work with states and the NMFS to conduct assessments of migratory fish stocks in state waters. Although assessments and statistics from FAO and NMFS provide only an imperfect characterization of the status of global and U.S. fisheries, the assessments—corroborated by many kinds of evidence—appear to provide a reasonably accurate description of the overall picture.

GLOBAL OVERVIEW

Fishing is an important source of food, recreation, community development, wealth, and cultural values in many countries. Although thousands of freshwater and marine fish and shellfish species are used globally, a relatively small number of these species provide the major fraction of the global marine catch. The 10 marine species that provided the greatest catch in 1993 accounted for 35 percent of the commercial marine catch and the top 20 species accounted for 46 percent of the global marine catch. Recent estimates indicate that the global first-sale revenues from fishery products are approximately $U.S. 95 billion annually and that fishery products account for about 20 percent of the animal protein consumed by humans. Fisheries provide direct and indirect employment to about 200 million people worldwide. Fisheries are especially important in developing countries, which increased their proportion of global catch from about 40 to 65 percent from 1973 to 1993. The net value of fishery products exported from developing countries totaled $16 billion in 1994, greater than the exports of coffee, bananas, rubber, tea, rice, and many other commodities that developing countries have traditionally relied on for foreign exchange.

Global marine fish production increased at an average rate of about 3.6 percent per year from 1950 to 1995, from about 18 million to about 91 million metric tons, including mariculture production. In the same period, the world's population increased from 2.5 billion to 5.7 billion people, an average annual increase of 1.8 percent. In 1995 total fish production (both freshwater and marine, both through culture and through fishing) was approximately 112 million t, of which marine landings accounted for approximately 84 million t. In 1996 the total production reached approximately 116 million t; the increase was due mainly to an increase in freshwater aquaculture production, mainly in China. The supply of fish and fish products for human consumption (including freshwater fish and aquaculture products) reached roughly 14 kg per person annually in 1995. By 1995, mariculture accounted for 6.7 million t, 7.4 percent of the total global marine fish yield; freshwater aquaculture provided 14.6 million t. Approximately 31.5 million t (28 percent) of world fish production was used for animal feed—including feed for mariculture—and other products that do not contribute directly to the human food supply in 1995.

In addition to fish that are caught and processed, a substantial number of fish and other organisms are caught and discarded—usually dead—at sea. Discards are a result of bycatch, which results because fishing gear and methods are not selective enough to catch only the target species, and of high grading, the discarding of smaller or less desirable fish in favour of larger or more desirable fish that are caught later. Some bycatch is retained, but the remainder is discarded when the species, size, quality, or condition of the fish reduce their value, or when fishery management regulations prohibit their retention. Alverson *et al.* (1994) estimated that commercial marine fisheries around the world discarded an estimated 27 million t of Non-target animals in the early 1990s, an additional biomass about one-third as large as total landings.

In addition to fish mortality caused by landings and discards, fishing can cause additional *unaccounted* mortality. Potential causes of unaccounted mortality include illegal or misreported landings; escapement or avoidance mortality that occurs when fish are injured by fishing gear but are not captured; and ghost fishing mortality, caused by lost gear (*e.g.*, traps and gillnets) that continues to catch fish. The magnitude of unaccounted mortality is unknown but may be high for some fisheries. For example, the Scottish Fishermen's Federation estimated illegal or misreported landings, probably of groundfish and crustaceans, to be 100 to 200 percent of the reported catch. Myers *et al.* (1997) concluded that discards of young undersized fish were an important reason that the fishing mortality of northern cod off Canada's maritime provinces was consistently underestimated in the 1980s, leading to the overfishing that caused the collapse of the fishery. Although these examples are not necessarily representative of all fisheries, they show that the total mortality resulting from fishing can easily be underestimated. The FAO (1994a) and Garcia and Newton (1997) have concluded that the relatively stable catches of the early 1990s indicated that capture fisheries are near, or have reached, their sustainable limit based on existing fishing techniques and market systems. The increase in catch between 1950 and 1994, occurred because of steadily increasing demand for fishery products, resulting in increased fishing capacity and effort. Fisheries were developed or expanded on formerly less-exploited or unexploited species and populations. While global assessments indicate that there are still opportunities to expand some fisheries, most are fully exploited or beyond, based on single-species considerations. Some have been so depleted that they are producing much less than their long-term potentials. Global fishing capacity is much greater than needed for sustainable marine fisheries, again based on single-species considerations.

More than 25 years ago, Gulland (1972) estimated that the potential sustainable yield from traditional fishery resource species (excluding Antarctic krill and oceanic mesopelagic fishes) was about 100 million t, with a practical limit (due to imperfect management and multispecies interactions) of about 80 million t, similar to catches that have been achieved in recent years and to recent FAO assessments. Maximum potential global marine fisheries yield has also been estimated by Schaefer (200 million t, 1965), Ryther (100 million t, 1969), Idyll 400–700 million t, 1978), Houde and Rutherford (more than 300 million t from all marine ecosystems, 1993), and others. Based on those estimates, one might conclude that we have not yet reached maximum global fisheries yield. However, most of the highest estimates—admittedly upper limits in some cases—were made using unrealistic assumptions about food-web structure, effects of bycatch, feedback effects of fishing on other fish populations and marine ecosystems, and the technical and economic feasibility of new fisheries.

Limits to Global Production

Three kinds of information suggest that marine fish catch is near, at, or above its maximum sustainable level: estimates of theoretical limits imposed by available primary production, information on the degree of utilization of fish populations, and information on the catch per ton of fishing vessel.

Food-Web Limitations

The capacity of the ocean to produce fish is limited in part by the amount of marine phytoplankton produced annually. Fishery landings tend to be higher from ecosystems with higher levels of primary production, especially marine areas characterized by fronts, convergence, and upwelling areas. Satellite and in situ measurements of phytoplankton concentrations and in situ measurements of nutrients, water temperature, irradiance, and primary production allow estimates of the primary production of the global ocean, as well as regional estimates. An upper limit to the ocean's potential fisheries yield has been estimated many times by applying knowledge of the amount and location of global primary production, trophic level of the catch, and the transfer efficiency of biomass among trophic levels.

Pauly and Christensen (1995) used global catch data—which they divided according to trophic level—to estimate the flow of carbon up through the trophic levels of global marine ecosystems. They estimated that the transfer efficiency between trophic levels was about 10 percent, and concluded that about one-quarter to one-third of total primary production in coastal and continental shelf waters is needed to support recorded landings plus discards.

Houde and Rutherford (1993) used relationships between catches and primary production and between fish production and primary production to estimate a partitioned global fisheries production for estuaries, coastal zones, and upwelling areas. They estimated a total global fisheries production in those ecosystems of 543 million t, from which 111 million t might be removed as yield. (Note that Houde and Rutherford's estimate of the total potential yield of more than 300 million t was based on an estimated production of more than 1,300 million t in all marine ecosystems.

They considered open-ocean production to be technologically difficult to use.) These estimates suggest that landings are near or beyond their sustainable limit, particularly if fish production lost as discards and unaccounted mortality are considered. Another line of evidence suggesting that global marine catch might not increase, even by fishing at progressively lower trophic levels, is provided by analyzing changes in the mean trophic level of marine fishery landings.

The average trophic level of fish catches from the 1950s for the northeast Atlantic and from the 1970s for the northwest Atlantic. This reflects a decrease in the proportion of long-lived carnivores in the catch relative to shorter-lived smaller pelages and invertebrates.

Fishing down the food web, while overfishing higher trophic forms, does not necessarily lead to increased total catches. As fishing takes animals lower in the food web, an increasing portion of the total catch may consist of animals for which there are no current markets or that are so diffuse that the cost of their capture does not warrant the expense (*e.g.*, some large zooplankton species).

In addition, the loss of predators (*i.e.*, animals higher in the food web) can lead to an increase in competitors of the target species. The average trophic level of landed species can drop rapidly as catches of top predators or decline as observed in most other FAO areas analyzed in this fashion.

CURRENT PROBLEMS IN THE MANAGEMENT OF MARINE FISHERIES

The management of commercial fisheries clearly requires a good scientific understanding of the behaviour of the exploited stock or stocks. The science that is used to assess commercially exploited species is still dominated by the population models developed by Beverton and Holt for single-species assessments some 50 years ago. The availability of substantial computing power has meant that sophisticated estimation methods can be used, and an appreciation of the way in which fish stocks respond to environmental variability is readily incorporated in scientific advice. Calls for a more ecosystem-orientated approach have been voiced for some while, but the paucity of data and the demands of multiparameterised multispecies models means that most ecosystem considerations in practical stock assessment tend to be *ad hoc* manipulations of the single-species approach. What has developed is a realisation that effective management requires an understanding of how the fishery system is performing. The most commonly used reference points are those relating to the size of the stock itself and the fishing mortality that will result in these stock sizes, given existing relationships between the stock, recruitment, natural mortality, and growth.

A typical "target reference point" is the biomass necessary to produce maximum sustainable yield. However, such targets do not explicitly recognise threats to the stock. To address this issue, stock size "limit reference points" are usually defined or interpreted as the stock biomass below which recruitment becomes substantially reduced. Clearly, it is important to avoid situations where the stock is at or below this level. Accordingly, management should aim to have as a target a level of stock size that carries a low risk (allowing for scientific uncertainty) of the stock dropping below the limit reference point. This could mean having a target level of fishing mortality that provides stock sizes above BMSY.

UNDERSTANDING FISHERIES MANAGEMENT

Competent scientific advice based on appropriate data is far from ubiquitous in the fisheries world, and even in ideal situations, fisheries management has often been unsuccessful. The success of a management system is often defined in terms of biological, economic, social, and political objectives. Clearly, economic and social objectives will not be met while a stock is in such a depleted state that the long-term sustainability of the fishery is threatened, but equally, biological objectives are unlikely to be met without consideration being given to economic and social objectives.

Hence, we argue that an understanding of the fishery management process can only come from analysing the capacity and incentives of the two key stakeholders: the fishing community and the management authority. This is not to belittle the importance of other stakeholders, such as recreational fishers and environmental groups, who have important roles in the management of certain fisheries. Where management is weak or non-existent, the economic factors underlying overfishing in commercial fisheries have been generally understood since the 1950s. In short, when multiple fishers compete to catch fish from a given population, each fisher maximizes his net income by continuing to fish as long as the value of his catch exceeds the cost of catching it. An equilibrium, called the bionomic equilibrium, is reached only when fishing has reduced the fish population to a level at which catch rates are barely sufficient to cover the costs of fishing.

The population is then maintained at this level through biological processes of natural growth and reproduction. Thus, if the price:cost ratio is high, the bionomic equilibrium will result in a low stock of fish, and hence a low annual catch level; two characteristic features of overfishing. In addition, the so-called economic rents (total revenue minus total costs) from the fishery will equilibrate at zero, resulting in minimal overall economic efficiency.

Many management authorities seek to meet their objectives by setting output controls in terms of a total allowable catch (TAC) for the year and closing the fishery when the year's cumulative catch has reached the TAC. Restrictions on fishing gear, fishing season, and fishing areas, as a supplement to the TAC, may also be imposed. If the TAC is correctly specified and enforced, this method should maintain a stock level well above that of bionomic equilibrium. If, however, the TAC and the science behind it are not respected by fishermen and not adequately enforced by authorities, widespread illegal fishing can occur.

A recent example is in the eastern Baltic cod fishery where illegal fishing contributes to true catches being some 35 to 40 per cent higher than reported. Efficient enforcement can be difficult. Put simply, fishers will be deterred from breaking fishing regulations if their expected loss from detection and successful prosecution exceed their expected gain. In many fisheries, the probability of detection of illegal activity and the penalties are not sufficiently high to act as a disincentive.

Strong management can ensure that biological targets are met, but it is essential that regulations are enforceable, and this has often proved to be difficult. Less-than-perfect enforcement can lead to illegal fishing, poor scientific data, and a failure to meet biological targets. Input measures, such as limiting the number of vessels or restricting available season length, are usually more easily enforceable than output measures such as TAC.

However, control via input measures is vulnerable to effort creep, whereby operators increase the fishing power of their vessels through technical means.

Nevertheless, monitoring of vessel performance over time and adjusting the allowable level of effort have allowed successful effort control to be implemented.

Overcapacity

Simplistically, it would seem that positive economic rents should also emerge in a TAC-regulated fishery. In reality, many TAC-regulated fisheries have experienced an unexpected increase in fishing capacity, as additional vessels enter the fishery in response to (temporarily) positive rents. Economic models then predict a regulated bionomic equilibrium, in which economic rents (net of fixed costs) again equilibrate at, or near, zero.

This situation currently exists in many of the world's regulated fisheries; overcapacity of fishing fleets is widely perceived as a major impediment to achieving economically productive fisheries. It is thus ironic that such overcapacity is usually generated by the management system itself, although it can also result from high profitability during the initial phase of a newly developing fishery.

Overcapacity is widely recognised as a major problem affecting world fisheries. With its attendant social and economic problems, overcapacity can, via the political process, lead to the erosion of management control. It is also understood to be one of the results of subsidising fisheries, which even today is estimated to be several tens of billion U.S., dollars per year. Such subsidies directly undermine the sustainability of fisheries because they lead to a bioeconomic equilibrium with high levels of fishing and low stock size. In several fisheries, government funds have been used to buy out excess fishing capacity. For various reasons, such buyback programmes have been less effective than expected.

First, often only the least efficient vessels are bought up, leaving total fishing capacity largely intact. Second, the buyback programme by itself does not remove the economic incentives underlying overcapacity, which tends to increase once the buybacks are completed.

Thus, the underlying cause of the dual crisis of overfishing and overcapacity, as well as other undesirable outcomes, such as habitat destruction and incidental kills of untargeted species, can be found in the economic incentives of fishers who compete for their annual catches. These incentives are not affected by management strategies that retain the competition between fishers for a common-pool resource.

Perhaps the most important development in fisheries management over the past 20 years has been the recognition of this fact and the introduction of rights-based management in several regimes. Indeed, it has been argued that of the tools at the disposal of managers, more emphasis needs to be placed on incentive-based approaches that better specify community and individual harvest or territorial rights, in additional to public research, monitoring, and effective administrative oversight.

Transferable Quotas

An alternative management strategy based on individually allocated transferable annual catch quotas (ITQs, or individual transferable quotas) is now in effect in several fishing nations, including Australia, New Zealand, Iceland, Canada, and Namibia. A well-organised rights system alters the economic incentives of fishers, who no longer compete for their catches, so that highly competitive fishing no longer takes place.

The guarantee to fishers of a certain proportion of the catch allows them to make rational economic choices about where and when they catch fish. An ITQ system goes further, allowing the industry to settle on a fleet capacity that optimises individual economic yield to vessels or cooperatives, although this of course can still be distorted by inappropriate subsidies.

In addition, ITQ fishers may often be expected to favour management actions that protect and enhance fish populations, because the value of a quota share increases as stocks become more abundant. Problems that may arise, such as misreporting or high-grading of catches, have been successfully countered by the use of observers, required by the management system but paid for by the industry; observers are used extensively in the U.S., Pacific fisheries, Australia, and New Zealand.

Experience with ITQ systems shows that many fishers willingly support and adhere to conservative management strategies and may also avoid fishing practices that endanger habitat or threaten other species, so long as they are guaranteed long-term rights. But this does not mean that enforcement and scientific monitoring are unnecessary in ITQ systems; both are essential unless catch levels are set at precautionary low levels. It is thus unsurprising that the two countries with perhaps the most fully developed ITQ systems, New Zealand and Iceland, have some of the highest costs of management per fishing vessel.

Several authors have pointed to instances of successful fisheries management in both the developed and developing world. Among their conclusions are that incentive structure, institutional capacity, and participation of stakeholders are of key importance. However, in some studies, a rights-based approach is seen as the primary mechanism to deliver this, whereas in others, severe top-down controls with very limited participation of fishing communities in the management process are advocated. We argue here that a necessary condition for successful management contains all these elements: a competent management authority able to set and enforce regulations and monitor the status of the stock, together with some form of rights-based allocation to fishing operators (either collectively or individually) to avoid the situation where overcapacity produces economic hardship and erodes management capacity.

Evidence from Fisheries Performance

Reviews of successful fishery management are of necessity specific to individual fisheries and sometimes anecdotal. However, in some large areas, a combination of strong state governance and wealth, substantial scientific activity, and different types of fishery management offer the opportunity of some comparison between different types of fishery management that goes beyond the anecdotal.

Detailed data on the status of different fisheries are published for U.S., Northeast Atlantic, Australian, and New Zealand fisheries. The approach to management taken by these authorities is varied. New Zealand has the most developed and widespread application of individual user rights (ITQs), which have been in place from 1986 and have spawned other developments such as collaborative and alternative research by stakeholders ITQs are present also in some Australian fisheries, a very few U.S., fisheries, and some Northeast Atlantic fisheries. (The Faroes, Norway, Iceland, and United Kingdom have rights-based systems, and some fleets, notably the Dutch flatfish and Spanish Grand Sole fleets, are also managed via ITQ.)

If a broad view is taken of these management areas, the evidence for the positive benefits of ITQs in supporting sustainable resource use is mixed. Only 15 per cent of New Zealand's stocks within the quota management system, for which the stock status is known, are substantially below the target reference level.

For other administrations, the percentage of stocks that are below the limit reference level, out of the total number of stocks for which the status is currently known, is 19 per cent for Northeast Atlantic fisheries managed by non-European Union (EU) administrations (Iceland, Faroes, and Norway), 25 per cent for federally managed U.S., fisheries, 30 per cent for Northeast Atlantic fisheries managed primarily by the EU, and 40 per cent for Australian Commonwealth fisheries.

Even within the United States, there are very large regional differences: 40 per cent of major Fish Stock Sustainability Index (FSSI) stocks managed by the New England and Mid-Atlantic Fishery Management Councils, for which the stock status is known, are overfished, and 30 per cent are subject to overfishing. By contrast, only 13 per cent of FSSI stocks managed by the Pacific, West Pacific, and North Pacific Management Councils are overfished, with 6 per cent suffering from overfishing.

Only the United States has seen an improvement in performance over the past several years; in 2000, 38 per cent of U.S., stocks for which the status was known were classified as overfished. All other areas have experienced some increases in the number of overfished stocks in the past decade, although the increases in New Zealand have been very small and are offset by recoveries in some inshore stocks. However, these statistics disguise a quite dynamic situation within each region; for example, of 74 U.S., stocks requiring rebuilding, biomass is increasing in 48 per cent of them even if they have not yet achieved rebuilt status.

More detailed examination of the U.S., situation reveals that although ITQs are not generally applied, West Coast fisheries are managed by quota controls with fishing rights assigned to fishing companies or sectors, whereas in the northeast, fisheries are managed by a days-at-sea scheme and other effort controls. In terms of their current performance and the stock recovery required by the Magnusson-Stevens Act, West Coast management systems appear to be more effective than northeast coast systems: Only two of the 18 New England stocks that were overfished in 1995 have now recovered, compared to 4 of 9 stocks similarly categorised by the Pacific Fisheries Management Council. Clearly, non-ITQ management systems do not always fail to maintain sustainable stocks, and management systems using ITQs are not always successful.

The critical additional requirement appears to be a formally adopted management strategy with predefined rules for what to do in different circumstances. In New Zealand, in addition to an ITQ system, a formal harvest strategy embedded in the Fisheries Act (1996) means that rebuilding is statutorily required when the stock is below its target level. By contrast, the lack of a formally adopted harvest strategy in the Australian Southern and Eastern Scalefish and Shark fishery has led to an increase in the number of overfished stocks in that fishery over the past 10 years, despite operating with an ITQ system since 1992.

EU fisheries also lack a formal harvest strategy; although there is a commitment under the 2002 revision of the Common Fisheries Policy to develop multiannual management plans for all stocks, such plans are only currently defined for 17 of the 94 stocks that fall under EU management, and many of these have had to be negotiated during periods of stock collapse. The key problems, *i.e.*, the need to provide incentives to fishers to engage constructively in fisheries management and the need to have strong legal support for predefined harvest strategies, apply equally to management of stocks under national control and those in international waters under the control of Regional Fisheries Management Organisations.

To our knowledge, none of the latter currently allocate rights to individual fishers, and only a few have defined and tested effective harvest strategies. Allocation problems continue to beset these high-seas fisheries and influence compliance, data availability, and transparency.

The recovery of depleted fish stocks is a key issue and one to which most countries committed themselves in 2002 as part of the World Summit on

Sustainable Development. An effective reduction in fishing effort, the participation of fishers and other stakeholders in the science and decision-making process, and the biology of the species are important factors affecting successful recovery.

However, unless a harvest strategy is defined, with pre-agreed, legally binding decision rules requiring reductions of effort when stock sizes decline below limit reference points, most management authorities will still delay taking action to recover stocks. Some of this delay may arise from uncertainty in the science, but mostly it arises from an unwillingness to take decisions that will create hardship for fishers, and usually a delay will exacerbate stock decline. Ultimately, the most successful management approaches are likely to combine rights-based systems, creating incentives for fishers to operate efficiently and with long-term sustainability in mind, with a strong legal structure that requires the development of pre-agreed harvest strategies and decision rules that are triggered and adhered to as reference points are passed. As indicated earlier, an adequate control of fishing activities is also necessary.

ADDRESSING ECOSYSTEM-BASED MANAGEMENT

In recent years, there have been many calls for much wider use of Marine Protected Areas to address the need for ecosystem-based management. We see these as a useful part of fishery regulation, but they are not a universal solution because unless the basic issues of capacity, regulation, and rights are solved, protected areas will simply displace the problem elsewhere. In this we concur with recent reviews that emphasise the primary importance of conventional measures to control fishing mortality and the secondary but essential role that marine protected areas (or other area-based management, such as local prohibitions on particular gears such as bottom trawls) have in dealing with specific issues of ecosystem conservation, such as bycatch and habitat damage.

The simple creation of rights-based incentives does not automatically deal with ecosystem problems, because fishers have little incentive to minimize bycatch or habitat damage that does not affect their target species. An interesting recent development is the creation of additional incentives for fishers through market measures, such as the creation of sustainable fisheries certification schemes and pressure from environmental non-governmental organisations for responsible fisheries.

Fishers have a major incentive to improve fisheries to satisfy certification conditions, and so far most of the conditions raised in Marine Stewardship Council certifications have concerned the ecosystem effects of fishing, often related to quantifying and reducing deaths of bycatch species and damage to habitat. Even in the statistics documented for some of those states with appreciable management capacity, what is striking is for how many stocks, the status is uncertain or not determined. In the United States, the stock status of 30 per cent of the 230 major (FSSI) stocks and stock complexes was undetermined in 2006; in Australia (48 per cent), New Zealand (78 per cent), and the Northeast Atlantic (61 per cent), the numbers are even higher.

Given the problems that most authorities have in deriving reliable quantitative assessments of their stocks of major commercial importance, the large numbers of small, commercially unimportant stocks present in most areas, usually as bycatch, cannot realistically be assessed. Under a comprehensive ecosystem approach, risk assessment methodologies should be used to identify those bycatch species in need of special measures, and monitoring programmes, for instance using scientific observers, need to be implemented to monitor trends in all bycatch species.

The application of these approaches is in its infancy even in the most advanced management schemes; many simply respond by setting untested but hopefully precautionary effort or catch limits. These considerations apply even more strongly to fisheries operators in developing countries. In a situation of little or no management capacity, some form of bioeconomic equilibrium is the likely result, but in such cases the management priorities may be different. Indeed, high employment with relatively modest economic rent, as long as it is compatible with the sustainability of the resource, may be a perfectly legitimate management goal. In other cases, the development of Territorial Use Rights (TURFS) within local communities can lead to effective management control and rights-based operations, resulting in successful management.

There is no doubt that there is a major problem with the world's fisheries, and, despite serious attempts to improve management and to facilitate recovery of depleted stocks, the success has been limited. The key issue that we highlight in this review is that for successful management a dual approach is required, one in which authorities provide incentives for conservation based on fishers' rights and which is supported by strong management incorporating legally enforced and tested harvest strategies.

ADAPTIVE CAPACITY IN MARINE RESOURCE GOVERNANCE

Marine resource governance is a global challenge of increasing magnitude and importance. Depletion of fisheries resources, coastal habitat degradation, threats of global warming, food security at risk and rising fuel costs all contribute to the challenge. Small Island Developing States (SIDS) in the insular Caribbean are particularly vulnerable to natural and anthropogenic threats. Fishing and coastal communities and enterprises depend heavily on marine resources and ecosystems. Cross-scale and multi-level networks for marine resource governance may provide means for enhancing resilience and adaptive capacity in these complex social-ecological systems. Understanding how local, national and transboundary networks link actors at various levels and scales across the region in their quest to use and manage shared marine resources is key to developing policies and practices that enable good governance.

The nature of such networks is being investigated through cases in small-scale fisheries governance in order to determine what features, including collaborative processes, relate to resilience and the capacity to adapt. This research is a work in progress, the conceptual framework and preliminary findings of which are shared in this chapter, along with early ideas on how research outputs can be communicated and used to improve marine resource governance in the Caribbean. The opening sentences of the Caribbean Sea Ecosystem Assessment (CARSEA), a sub-global component of the Millennium Ecosystem Assessment (MA) proclaim that: "The peoples of the Caribbean are defined by the Sea whose shores they inhabit.

In the rich diversity of cultures and nations making up the region, the one uniting factor is the marine ecosystem on which each ultimately depends". Yet, Caribbean marine resources are in trouble, as is the case in many coastal and marine areas worldwide. Depletion of fisheries resources, coastal habitat degradation, global warming, food security at risk and rising fuel costs all contribute to the challenges faced by the Small Island Developing States (SIDS) in the insular Caribbean. These places and their people, characterised by coastal communities and enterprises (mainly fishing and tourism) heavily dependent on marine ecosystems, are particularly vulnerable to both natural and anthropogenic threats to life and livelihood. Member states of the Caribbean Community (CARICOM) are among the most numerous, dependent and vulnerable in the region. Looking closer at fisheries, throughout CARICOM there is a need to develop alternatives to conventional fisheries management, paying particular attention to people-centred approaches and governance.

On a global scale conventional science-centred approaches to fisheries management have not performed well, and some scholars have deemed them to be failures. As a result, fisheries have long been said to be in crisis despite recent attempts to incorporate Ecosystem Approaches to Fisheries (EAF) and add human dimensions to this. Greater attention not only to ecosystem approaches to marine science, but to governance as well, is needed to address situations that threaten Caribbean social-ecological systems.

This is the convergent conclusion that can be drawn from several sources as summarised in CARSEA and the Caribbean Large Marine Ecosystem (CLME) project. These reports, and this chapter, focus upon the use of living marine resources and governance of small-scale fisheries within the context of ecosystem approaches.

Fisheries management and integrated coastal management are still relatively new in CARICOM. Before conventional approaches firmly take root, the opportunity exists to investigate, test and propose approaches based on concepts and models drawn from complex adaptive system (CAS) and social-ecological system (SES) perspectives to determine whether these new alternative directions offer greater promise of success than the conventional. In the fisheries literature these perspectives have not yet become mainstream, but there is a growing interest in their theory and practical application, especially with regard to marine resource governance in small-scale fisheries.

At the core of the new perspectives are concepts and practices still unfamiliar to many policy decision-makers, management authorities, resource users and other stakeholders engaged in marine resource governance. These core concepts include social networks, resilience and adaptive capacity. Put simply, they are related in that network structures and relationships provide a useful way to represent complex SES, resilience can be a desirable feature of SES, and adaptive capacity is one of the means of achieving and sustaining resilience in SES.

We are engaged in research and development to put some of this thinking into practice. In this chapter we examine small-scale fisheries governance in the CARICOM region with an emphasis on the role of user groups, particularly fisher folk organisations (FFOs). This considers what and how multi-stakeholder collaborative planning that links fisheries authorities and fisher folk at national and regional scales with university researchers can contribute to resilience thinking and action.

At this early stage of engagement much depends upon communication, so we also pay attention to how information exchange is helping to shape this ongoing experiment.

CARIBBEAN MARINE RESOURCE GOVERNANCE

The Caribbean is culturally, ecologically, geographically, institutionally, politically, socially and otherwise diverse and complex. Considering only the 15 CARICOM Member States scattered across the region does not significantly reduce complexity.

From an ecological perspective, one of the main contributors to complexity is that most marine resources, both exploited and non-exploited, are shared across several maritime jurisdictions at various points in their life cycles. From a social system perspective, there is also considerable interaction of people and issues across jurisdictions.

Marine resource governance in the region must therefore be transboundary, crossing the Caribbean at multiple scales and levels. Some transboundary linkages are not commonly known, and consequently the need for collaboration in governance is poorly appreciated. This SES cross-scale complexity and limited collaboration in governance results in what some authors refer to as a maze of regional and sub-regional institutions, such that the many overlapping, competing arrangements for various aspects of marine governance do not provide a coherent governance framework. This is a critical problem, especially given the limited state and civil society capacity for fisheries management within the region, and the scarcity of fisheries officers trained in social sciences and related skills.

To reduce the problems of scale mis-match, while not discounting the larger Wider Caribbean complexity and the several other regional initiatives that concern marine resource governance such as the CLME project, we concentrate on arrangements linked to the Caribbean Regional Fisheries Mechanism (CRFM) established in 2002 by intergovernmental agreement as a body of CARICOM.

The early evolution of the CRFM. The mission of the CRFM is "to promote sustainable use of fisheries and aquaculture resources in and among member states by the development, management and conservation of these resources in collaboration with stakeholders to benefit the people of the Caribbean region" (emphasis mine). The CRFM is the core of a complex interactive network of a wide variety of stakeholders in fisheries. Three organs together make up the CRFM: (1) the Ministerial Council or Body; (2) the Caribbean Fisheries Forum; and (3) the CRFM Secretariat.

The former makes policy decisions and the latter provides technical and administrative support. However, for fisheries stakeholders, most of the action should take place in the Forum.

The Forum is the main technical and scientific decision-making arena, but it may also be the most important in terms of fisheries governance unless its advice is questioned or overridden by the Ministerial Council. The Forum comprises full members, associate members, and observers. They says that "Observers may include, but are not limited to, intergovernmental fisheries organisations, fisher folk organisations and other NGOs, and universities". The CRFM Secretariat is the secretariat of the Forum.

Based on the foregoing, the Forum is an attractive target for FFOs and private sector fishing industry groups that wish to get into the CARICOM policy domain through the route of being observers. It is also possible that they may be included in member state delegations as has been the case with the Suriname Seafood Association, but this is less likely in most other member states. Apart from the Suriname example, FFOs have been absent from the Forum and none are accredited to it as observers on their own.

Another attraction of the Forum is that in 2003 it formed an open-ended working group to take on the task, directed by the CARICOM Heads of Government, to establish a Common Fisheries Policy and Regime (CFPR). The CFPR is still under negotiation. What its final form and function will eventually be is presently unpredictable.

However, the CFPR may become one of the main instruments for shaping and implementing fisheries governance within CARICOM and perhaps beyond.

Early drafts of the negotiation text suggest that the Forum contemplates having, through the CFPR and otherwise, a wide range of decision-making responsibility and authority. Matters for its consideration may range from all aspects of harvest operations (including perhaps access and allocation), to postharvest and trade, marine aquaculture, research, enforcement, dispute settlement and more. One of the main reasons that FFOs have been unable to engage the CRFM in marine resource governance is that in most countries they lack the capacity at both local and national levels, and they have not yet formed a regional level organisation. McConney describes how FFOs were introduced to many locations in the English-speaking Caribbean (now the Member States of CARICOM) during the British colonial period, often in the early 1960s and 1970s. The histories of how they rose, and often fell, at different times in different places vary, but common inter-related features of their state-directed formation typically include:

- Being part of the movement towards independence
- Use for political empowerment of "the small man"
- A means to alleviate poverty and encourage saving
- For pooling money to improve commercial business
- An emphasis on boat owners rather than fishers
- Channelling of government subsidies for fishing
- Efficient supply of inputs for fishery development
- Getting into bigger, better and costlier boats
- Seafood processing for food security and export
- Starting with savings societies and cooperatives

This illustrates a variety of reasons behind organising, and of different types of organisation (harvest, gear and fuel supply, processing, etc.). Early cooperatives failed in many places after only a few years for reasons including:

- Under-capitalisation (not enough funds to run them)
- Limited capacity (leadership, skills, resources, etc.)
- Inadequate support from government agencies·
  Membership bases that were too small to be viable
- Little succession planning to groom new leaders
- No follow-up to projects and technical assistance
- Short term planning rather than strategic planning
- Limited oversight and monitoring to give guidance
- Financial mismanagement ("who thief the money?")
- Regulatory agencies did not correct financial wrongs
- Connection to political movements that changed
- Inability to sustain collective action outside of crises
- Government programmes that competed with them

Although the development trajectories of FFOs in CARICOM are poorly documented, through oral history we are able to look back on these past experiences and learn from them. Yet, learning from past successes and failures is not adequate preparation for the present and future if situations have changed. How things have changed varies by location and the diverse nature of Caribbean fisheries SES, but the change that most concerns us here is the recent emergence of the CRFM and the opportunity for involving FFOs in cross-scale governance at integrated multiple levels whereas previous efforts were mainly local or national.

Recognising this opportunity, the CRFM initiated a project on the 'Development of a Caribbean Network of Fisher Folk Organisations' (emphasis mine) in September 2006, implemented through its Secretariat. The demand for such a project was based on the results of a regional needs assessment and subsequent regional meetings in 2004 and 2005 that involved FFO representatives and recommended the formation of a regional network of national fisher folk organisations (NFOs). Strengthening FFO capacities was to be aimed at addressing:

- Lack of critical mass of NFOs for launching the regional network
- Fisher folk organisations' weak management skills
- Insufficient access to relevant information by fisher folk

The 2005 meeting produced a 'Strategy and Medium Term Action Plan for the Institutional Strengthening of Regional Fisher Folk Organisations-2006 to 2010' to address the points above. It is through this bottom-up collaborative planning process that the project was designed. The specific purpose of the project is to have "institutional capacities of fisher folk organisations developed at the regional, national and community levels".

In the two years since its inception in 2006 the 'FFO project', as it has come to be known, has held three regional training workshops, created a directory of fisheries stakeholders, produced bi-monthly newsletters, spawned a fisher folk lead coordinating unit (CU), sponsored exchange visits of fisher folk leaders and supported the implementation of a work plan to establish the networked regional fisher folk organisation (RFO). The structure and functions of the latter are still being worked upon.

Researchers at the University of the West Indies (UWI) Centre for Resource Management and Environmental Studies (CERMES) have been involved in the FFO project as a partnership with the CERMES project on 'Marine Resource Governance in the Eastern Caribbean' (MarGov) Project.

The latter is participatory action research aimed at developing capacity for marine resource governance. The goal of MarGov is to understand marine resource governance related to small-scale fisheries and coastal management in the eastern Caribbean using CAS and SES concepts. Before we examine the relevance of the FFO project to this, the use of some concepts is briefly explained.

NETWORKS, RESILIENCE AND ADAPTIVE CAPACITY

Networks consist here of the nodes and ties that represent the actors and relationships among actors in a fisheries SES. Relationships may embody trust, power, information, financial or technical assistance and more. It has been argued that studies on resilience in SES lack a clear analytical framework, and that a network perspective with emphasis on dynamic, mixed systems might be useful. This is because network analysis focuses on the structure of interactions between SES components and the ways in which nodes and ties affect the performance of systems at multiple scales. Network analysis has been applied to both social systems and ecological systems, combining qualitative and quantitative information, but less often to integrated SES and governance.

Further suggest that there is no 'right' depiction of SES networks, only models and maps that are more or less useful from a particular perspective or for analysing a specific problem. They also warn that network architecture or structure is of little value unless the flows through the network are also known. Finally, they conclude that there is no simple relationship between network properties and 'good' or 'bad' resilience. For example, a dense network may be good for the rapid diffusion of a beneficial innovation, but it may also spread unwanted practices such as irresponsible fishing and constrain learning and adaptation by individuals trapped in close-knit cliques that are resistant to adaptive change.

Resilience is the capacity of a system to experience shocks or perturbations while retaining essentially the same function, structure, feedbacks, and therefore identity. The more resilient a system is, the larger the disturbance it can absorb without shifting into an alternate regime. Such a shift may be reversible or irreversible either absolutely or relative to the time scale of interest. Whether a particular regime is 'good' or 'bad' is largely a value judgement that varies with the experiences and perspective of the stakeholder. Resilience may not be advantageous if a fisheries regime is undesirable, so knowing what increases or reduces resilience is important for adaptive governance. Interventions that aim to alter resilience in SES soon confront power and governance issues.

Adaptive capacity reflects learning and the ability to experiment and foster innovative solutions in complex social and ecological systems. To learn and innovate in the process of self-organisation and adaptation systems must be open to and tolerant of failure.

Yet, in the political reality of most governance arrangements, experimental policy (with the risk of failure) is seen as being prohibitively costly in many respects.

However, adaptive network governance and strengthening adaptive capacity to manage resilience are critical to sustainable development. Identify critical components of adaptive capacity and resilience that interact across temporal and spatial scales during periods of change and reorganisation. These include learning to live with change and uncertainty; nurturing diversity for resilience; combining different types of knowledge for learning; and creating opportunity for self-organisation towards achieving social-ecological sustainability.

MARINE FISHERIES RESOURCES

The marine fishery resources of the country's EEZ stand assessed at 3.93 million metric tonnes as per the latest update of 2000. This resource is distributed in inshore (58 per cent), off shore (34.9 per cent) and deep sea (7 per cent) waters. The major share of this resource is demersal (2.02 million tonnes) followed by 1.67 million tonnes of pelagic and 0.24 million tonnes of oceanic resources. The estimates also points to the fact that there is scope for further augmenting the marine fish production by about 1.2 million tonnes if fishing is carried out deploying resource-specific vessels, mainly in the oceanic region. Another phenomenon noticed is the depletion of resources in the coastal sector, which is either species specific or location-specific, both resulting from unsustainable fishing pressure. The policy therefore underscores the need for a departure from the open access concept in the territorial waters besides putting in place stringent management regimes. Promoting exploitation in the deep sea and oceanic waters would be another approach for reducing fishing pressure in the traditional fishing areas.

HARVESTING OF MARINE FISH RESOURCES

Harvesting of marine fish resources is categorised into three levels *viz.*, (i) subsistence fishing (ii) small-scale fishing and (iii) industrial fishing. The policy advocates protection, consideration and encouragement of subsistence level fishermen and technology transfer to small scale sector and infrastructure support to industrial sector.

There would be exclusive area in terms of depth and (or) distance earmarked for non-mechanised (non-motorised) traditional craft. An area beyond this would be demarcated for mechanised and motorised craft. The area for deep sea fishing vessels including all boats above 20 m OAL and fitted with inboard engine and having chilled or refrigerated fish hold would be beyond the limits prescribed for the other two categories of vessels.

Within the territorial waters, the demarcation of area for traditional, motorised and small-mechanised fishing vessels is the purview of the coastal state/union territory. Efforts would be made to harmonise the demarcation of reserved areas to the maximum extent possible so that a uniform pattern is followed in all coastal states/UT's.

Encouragement to subsistence level fishermen would include scheme to motorise the traditional craft and also providing better material and technology for their traditional craft. The country has a very large fleet of traditional craft. Motorisation of the entire fleet may make fishing un-sustainable. The motorised craft with their operational limit would end up in overcrowding whereby excerting too much fishing pressure in a limited area. The Policy therefore envisages motorisation of about 50 per cent of traditional craft allowing the remaining to carry on subsistence fishing in the near shore waters.

The small-mechanised sector would be encouraged by providing incentives for acquisition of multi-day fishing units.

Deep sea vessels would be provided with infrastructure support in terms of landing and berthing facilities. The vessels, which are landing quality fish for export would be provided with suitable incentive as in other export oriented agri-ventures.

As the bulk of incremental catch to augment annual marine fish production has to come from deep-sea sector and beyond EEZ limit, the Government would encourage introduction of more resource specific vessels of above 20 m length.

Proposals for import of resource-specific fishing vessels by wholly Indian owned enterprises would be screened and approval accorded for such imports by a designated authority in accordance with well laid out norms. These additional fishing units in the deep-sea sector would be for tuna fishing and squid jigging. Special incentives would be provided for wholly Indian owned vessels for venturing into international waters and for concluding fishing arrangements with other nations under license etc.

Joint venture initiatives with specified equity norms for package proposals involving catching fish from the EEZ for processing at shore and export would be considered.

Fishing in Antarctic waters by Indian owned vessels or with equity participation or under license would be promoted by working out sustainable strategies. The principle of Code of Conduct for Responsible Fishing Operations would be incorporated into every component activity. Assessment of existing fishing capacity and plans for regulating or developing one or the other sectors of EEZ would be taken up.

The existing joint venture fishing vessels would be required to fully indigenise their operations as per the original proposal upon which permits were granted to these companies. No substitution for joint venture vessels would be permitted.

POST-HARVEST OPERATIONS

Total utilisation of harvested fish for food and non-food uses would be the central theme. Efforts would be made to fully comply with international requirements in post harvest care of catch so as to achieve highest standards in food safety. It would be also the concern of the Government to ensure that the post-harvest losses are minimized.

Implementation of international quality regimes for ensuring food safety in fish and fishery products would be carried out through the nodal agency. A regulatory body would ensure monitoring and verification of compliance. Existing domestic standards for fishery products and by-products would be harmonised with the International Standards so as to ensure quality of fish and fishery products for domestic consumption at par with global standards. Packaging and bar coding would be made mandatory for authorised sale of fish and fish products through registered outlets for ensuring food safety. Consumer rights would be given due attention in domestic trade of fish and fishery products. Hygiene in fishing harbour/pre-processing and processing centres would be streamlined through legislation.

RESOURCE MANAGEMENT

Exploitation of living resources within 50 metres depth zone is showing symptoms of depletion and in certain belts in the inshore waters it tends to cross optimum sustainable levels. The policy therefore advocates a stringent fishery management system to be in place.

Though the Marine Fishing Regulation Acts (MFRA's) of coastal states and UT's have adequate provisions for management of resources and fishing operations, it is often found falling short of effective implementation. This calls for a review of the situation and prescribing a fresh model bill on coastal fisheries development and management with a re-orientation on limited access in coastal marine sector through policy initiative, sound legislation and awareness creation. Construction and introduction of new fishing units cannot go unchecked any more. All existing boat-building yards shall be registered and construction of any new fishing unit will be after obtaining a license. Standards for fishing vessel construction, especially for those below 20m OAL need to be developed and control would be exercised through new legislation. Provisions would be made to comply with requirement of registration of vessels and Standards of Training, Certifications and Watch keeping of fishing vessel personnel.

There will be closed season on both the coasts, the duration of which would be decided by a designated authority. Such closed seasons shall be uniform for neighbouring states unless the geographic or climatic conditions warrant deviations. There would be strict ban on all types of destructive methods of fishing. The designated authority would be competent to declare any method as destructive after it is convinced so based on facts and data pertaining thereto.

Mesh sizes in different parts of the fishing gear would be regulated. Penalties would be fixed for violations of mesh regulations. The designated authority would, if found required doing so, decide the quota for different classes of fishing vessels in any region. Catching of juveniles and non-targeted species and discarding less preferred species once they are caught would be strictly prohibited through legislation.

Posting of observers on commercial fishing vessels and enforcing monitoring control and surveillance system (MCS) would be ensured. A resource enhancement programme will be taken up on priority.

This would include setting up of multi-species hatcheries for producing seed as required for sea ranching. Designating certain areas as marine sanctuaries and regulating capture of brood stock from these locations would be implemented. Open sea cage culture would be promoted to rear or fatten commercially important species of fishes. Fish aggregating devices would be promoted as a community based activity.

INFRASTRUCTURE DEVELOPMENT FOR MARINE FISHERIES

Development of infrastructure for marine fisheries is of vital importance and should have an integrated approach. The facilities would *inter alia* include jetties, landing centres, provision for fuel, water, ice, repairs to vessels and gear. The concept of hygienic post harvest handling of fish would also be woven into the project. The policies in this direction would be as follows:

- A master plan for the development of infrastructure for the next ten years would be drawn up.
- Alternatives to the present system of financing of the infrastructure projects by the centre and the state with cost sharing would be tried out. Build-Operate-Own and Build- Operate-Transfer systems through private sector initiative also would be explored.
- Management of most of the facilities already created calls for improvements in terms of internal resource generation, maintenance and upkeep. These issues would be subjected to a detailed study and suitable central legislation would be introduced if found necessary.

LEGISLATIVE SUPPORT

An enabling legal framework is an essential pre-requisite for proper management and control of fisheries sector. As at present the subject of fisheries is in the state list under article 21 of the Indian Constitution, management and control of coastal fisheries is vested with the maritime states and union territories. At the same time the Union Government carries out management and control of the fishing activities beyond territorial limits in the EEZ. Besides reviewing the existing legal frame work for regulating the fishing operations, introduction of additional legal instruments in areas such as operation of Indian flag vessels in the EEZ, introduction of new fishing units, ensuring conservation of resources, limited access fishery, fishery harbour management etc. would be resorted to.

In view of increase in the incidence of straying by small-mechanised boats into each other's territorial waters and consequent confiscation and arrest of crew, a mutually agreeable system will be brought in place with friendly neighbouring countries to have a lasting solution to the problem. Endorsing international laws and conventions in the marine fisheries sector and harmonising the national laws with the international ones wherever necessary with active participation in the regional fisheries management bodies and greater cooperation amongst countries in the region would be given due attention. Participation in the Regional Fisheries Management Bodies (RFM) should be given due consideration for greater co-operation amongst the neighbouring countries in the region.

Areas such as use of information technology, strengthening of database in marine fisheries, Human Resource Development, eco labeling of marine products, would also be paid needed attention.

MANAGEMENT OF RESERVOIR FISHERIES

The fish production from the reservoirs is low, emphasising the need for attention to shape and develop the reservoir fisheries from the survey and planning stage to achieve high rate of production and better returns for the fishermen, who represent the weaker section of the society. Majority of these water bodies are not scientifically managed. Only a handful has so far been harnessed on scientific lines, while the others are either half-heartedly managed or even not managed at all.

There are marked variations in the fishery management practices which are followed in various reservoirs within the country. Even though the reservoirs are owned by the Government or Corporate agencies in most of the states, their fishing rights and exploitation systems vary considerably. The fishing systems can be divided into the following broad categories: a). privately owned and managed reservoirs, b). Public water bodies, c). Community water bodies and d). Water bodies managed by the Government. After a scrutiny of the various management practices followed in the country, it is difficult to miss a common underlying spirit of the common property norm. Majority of Indian reservoirs are public properties where a fixed number of licensed fishermen make their living. The exceptions are the small reservoirs in some states like Karnataka and Uttar Pradesh, which are auctioned to private individuals on annual basis. The following steps are to be taken for the development of reservoir fisheries. Some of them are

- Pre-impoundment survey of the reservoir
- Removal of obstacles like tree stumps.
- Fish farm construction at the dam site for stocking of fish seed adequately.
- Organisation of fishermen co-operative society for harvesting the fish in the reservoir and to take up fish marketing.
- Implementation of conservation methods to prevent over exploitation and to prevent catching of small fish.

Pre - impoundment survey: Before formation of reservoir studies and investigations were taken up in the rivers connected to know the possible effects of fisheries. In case of the dams constructed, the dam constructing authorities should facilitate necessary fish ways for fish migration. The authorities also assess the fishermen living in the villages and their craft and tackles that are going under the submergence of the reservoir. The above measures are to be taken up before formation of the reservoir to assess and to plan for future development of fishery in the reservoir.

Removal of obstacles under submergence: After formation of the reservoir, it should facilitate easy fishing with modern fishing methods. To facilitate the above, the trees, buildings, boulders are to be removed either by mechanical means or manually.

Construction of fish seed farms at dam site: Some of the reservoirs have got river connection as such, they will have natural stocking of fish seed (auto stoking). The reservoirs, which will have a deep column of water, will greatly affect the breeding grounds of the fish. Further more the height of the dam is another major obstruction for the movement of the fish from the lower stretch of the river to upper stretches.

All these factors leads to low productivity, which drags the attention of many authorities towards the fish seed stocking every year in the reservoirs. Keeping in view of this, planning has to be done for establishment of a fish seed farm near the dam site.

The Department of fisheries, Govt. of Andhra Pradesh made an attempt to promoting the fishery wealth in the reservoirs, by constructing the fish seed farms at Thandva, Wyra, Kinnerasani, Palair, Nagarjuna Sagar, Nizam Sagar, Dindi, Jurala, Kadam, Araniyar, Bahuda, Somasila, Sriram Sagar and Lower Manair dam. Stocking of the fish seed in reservoirs: An adequate sized fish seed should be stocked in the reservoirs. The tender stages of fish seed like fry (22-25 mm), fingerlings (40-50 mm) are susceptible and are to be eaten by the carnivorous fishes. Hence the reservoirs to be stocked with an advanced fingerlings (100-120 mm). Stocking of reservoirs with fingerlings of economically important fast growing species to colonies all the diverse niches of the biotype is one of the necessary prerequisites in reservoir fishery management. This has proved to be a useful tool for developing fisheries potential of such small aquatic systems.

Agarwal recommended the following stocking rates in reservoirs.

- Large reservoirs (1000-5000 ha) - 500 fry/ha/yr.
- Medium reservoirs (100-1000 ha) - 1000 fry/ha/yr.
- Minor reservoirs (10-100 ha)-2000 fry/ha/yr.
- Small reservoirs (below 10 ha) - 10000 fry/ha/yr.

Srivastava recommended the following stocking rates of fingerlings for reservoirs in India.

- Large reservoirs (5000-10000 ha) - 200 fry/ha/yr.
- Medium reservoirs (1000-5000 ha) - 400 fry/ha/yr.
- Minor reservoirs (up to 1000 ha)-1000 fry/ha/yr.

IMPLEMENTATION OF CONSERVANCY METHODS TO PREVENT OVER EXPLOITATION

It has been observed that fishes of mature major carp fishes migrate to the rivers for breeding, where the fishermen lead to catch them and destroy their breeding grounds. The breeding grounds of these fishes are to be identified and necessary protective measures are to be taken up. It is also observed that during rainy season, the big fishes migrate towards the dam site along the canals and congregate at the gates, which has been catch by the fishermen folk. This is to be reduced.

The conservancy methods like mesh size regulation, observing the closed season for fishing, strict observation of prohibiting the unethical means of fishing are to be implemented strictly by the Government.

Species enhancement: Decline of indigenous fish stocks due to habitat loss, especially that caused by dam construction, is a universal phenomenon. Planting of economically important, fast growing fish from outside with a view to colonising all the diverse niches of the biotope for harvesting maximum sustainable crop from them is species enhancement. It can be just stocking of a new species of introductions. Here introduction means one time or repeated stocking of a species with the objective of establishing its naturalised populations. This widespread management practice has more relevance to larger water bodies where stocking and recapture on a sustainable basis is not feasible.

Introduction of exotics: In India, the fish transferred on trans-basin basis within the geographic boundaries of the country is not considered as exotic and there are no restrictions on them. Thus, catla is not regarded as exotic to peninsular rivers.

This is despite the fact that the peninsular rivers have habitats distinctly different from those of Ganga and Bhrahmaputra. Catla, rohu and mrigal have been stocked in the peninsular reservoirs for many decades now with varying results. In some of the south Indian reservoirs, they have established breeding populations. The hallmark of the Indian policy on introductions is the heavy dependence on Indian major carps.

There is evidence that the Gagetic major carps have affected the species diversity of peninsular cyprinids. The Indian policy on stocking reservoirs, though not very explicit, disallows the introduction of exotic species into the reservoirs. However, common carp is very popular in reservoirs of the northeast where it enjoys a favourable microclimate and a good market. Silver carp and grass carp are not normally encouraged to be stocked in Indian reservoirs, though they are stocked regularly in a few small reservoirs. The three exotic species brought in clandestinely by the fish farmers, bighead carp..Aorichthys nobilis, Oreochromis niloticus and African catfish Clarius gariepinus have not gained entry into the reservoir ecosystems. They still remain restricted to the culture systems.

Environment enhancement: The improvement of the nutrient status of water by the selective input of fertilizers is a very common management option adopted in intensive aquaculture. If similar environmental enhancement is adopted in small reservoirs, stocks can be maintained at levels higher than the natural carrying capacity of the environment.

However, scientific knowledge to guide the safe application of this type of enhancement and the methods to reverse the environmental degradation, if any is still inadequate. On account of all these, this is not a very common management tool. China is known to have used this instrument in a big way to augment production from small reservoirs. Cuba, taking a clue from China has tried manuring of small reservoirs using both organic and inorganic fertilizers. This is also practised selectively in the community water bodies of Thailand.

Fertilisation of reservoirs as a means to increase water productivity through abetting plankton growth has not received much attention in India. Multiple use of the water body and the resultant conflict of interests among the various water users are the main factors that prevent the use of this management option. Surprisingly, fertilisation has not been resorted to even in reservoirs which are not used for drinking water and other purposes. Documentation on fertilisation of reservoirs in India is scarce. Sreenivasan and Pillai attempted to improve the plankton productivity of Vidur reservoir by the application of super phosphate with highly encouraging results. As soon as the canal sluice was closed, 500 kg super phosphate with $P_2,0_5$ content of 16 to 20 per cent was applied in the reservoir when the water spread was 50 ha with a mean depth of 1.67 m.

As an immediate result of fertilisation, phosphate content of water increased from nil to 1.8 ppm and that of soil from 0.242 to 0.328 per cent. Similar improvements in organic carbon nitrogen have been reported from soil and water phases on account of fertilisation. Experiments were also conducted with urea in the same reservoir.

Fertilisation can play a key role in many small reservoirs of India which require correction of oligotrophic tendencies. A number of reservoirs in Madhya Pradesh the Northeast and the Western Ghats receiving drainage from poor catchments show low productivity, necessitating artificial fertilisation. Chinese experience in fertilising the small reservoirs for increasing productivity has been reassuring. Fertilizers are less effective in soft water with total alkalinity less than 20 ppm. Soft waters have inadequate carbon usually in the form of carbon dioxide and can often be enhanced by applying lime to low alkalinity impounded waters. The application of lime equivalent to 2,000 to 6.000 kg/ha calcium carbonate is generally sufficient to maintain total alkalinity above 20 ppm.

MODEL FOR MANAGEMENT OF RESERVOIRS

The concept of fishery-business model was developed by Desai in 1984. The main components of this model are macro-fisheries and micro - fishery systems.

Macro-fishery system again composed with fish supply system, fish production system and fish marketing system. The micro -fishery system contains fishery input sub system, fishery credit subsystem, fishery extension subsystem, fishery education sub system, fishery habitat sub system, fishery research subsystem, fishery regulation subsystem, fishery administration subsystem, fishery exploitation subsystem, fishery processing subsystem and fishery distribution sub system.

Natural Fishery in Reservoir

The following fishery wealth is available in the major reservoirs of Andhara Pradesh. The Fishes like Wallago attu; Mystus species; Heteropneustes fossilis; Clarias sp; Major carps like - Catla catla; Labeo rohita; Cirhinus mrigala and other carps like C. reba; Thynicthus sandkhol etc. The fresh water Eel, fresh water prawns, Murrels are also available in the reservoirs.

Planning Criteria

In keeping with the need for rapid assessment of the country's small reservoir resources, the following planning criteria are suggested for the resource assessment at the State level for the preparation of an inventory of such small reservoir ecosystems along with the estimates of their potential yields under the categories-a. Reservoirs which are best developed as capture fisheries, b. Reservoirs mostly of local interest having significant potential for fish culture; and c.Reservoirs intermediate in size and potential yield.

These under-utilised fishery resources offer immense scope and potential for generating additional national income of the order of more than ₹ 100 crores per year and providing additional employment to lakhs of fishermen and others through fishing, handling, transport, marketing and ancillary industries. A systematic and integrated approach towards scientific studies and planning criteria for undertaking fish culture in small reservoirs should be so directed as to have an understanding of the following factors.

- The reservoir morphometry and water residence time.
- The physico-chemical characteristics of water and soil.

- The animal and plant inhabitants.
- Growth rate of commercially important fish species, and
- The relation between the inhabitants and the physico-chemical aspects of the environment in terms of population and community dynamics.

It is felt that under the prevailing socio-economic conditions, such short-range studies undertaken for small reservoirs would provide a rapid assessment of their fisheries potential to take up fish culture in them.

FISHERY MANAGEMENT AND WILDLIFE RESOURCES

The fish and wildlife resources department at the University of Idaho is among the finest in the nation. With a Bachelor of Science in Fishery Resources, you will be equipped to develop solutions to some of the world's most pressing concerns in fish ecology and management. Commercial landings (edible and industrial) by U.S., fishermen at ports in all the fishing states were a record 8.5 billion pounds (3.8 million metric tons) valued at $3.2 billion in 1989. This was an increase of 1.3 billion pounds (576,300 metric tons) in quantity, but a decrease of $281.8 million in value, compared with 1988. The total import value of edible fishery products was $5.5 billion in 1989, based on a record quantity of 3.2 billion pounds. Imports of non-edible (industrial) products set a record in 1989, with products valued at $4.1 billion, an increase of $676.1 million compared with 1988.

The trade deficit in fishery products has not declined. The dollar value of imports was higher in 1989 than in the previous year. Canada is still the largest exporter to the United States, sending in more than 700 million pounds of fishery products in 1988. Ecuador was ranked second and Mexico third. Whereas Canada ships finfish products, shrimp is the primary commodity exported by Ecuador and Mexico. Imports from Thailand and China are both increasing due to rising shrimp production from their expanding aquaculture systems.

On a worldwide basis, aquaculture is becoming a major new factor in seafood production. The cultivation of high-value species, popular in the U.S., market, is a major factor in import sourcing. China, for example, along with other Asian nations, is replacing South and Central American countries as a major shrimp supplier to the United States. Aquaculture is expected to determine much of the future fisheries growth, because wild stocks are nearing full utilisation. The total export value of edible and non-edible fishery products of domestic origin was a record $4.7 billion in 1989, an increase of $2.4 billion compared with 1988. The United States exported 1.4 billion pounds of edible products valued at $2.3 billion, compared with 1.1 billion pounds at $2.2 billion exported in 1988. Exports of non-edible products were valued at $2.4 billion. Japan continues to be America's best export customer.

Over 700 million pounds of seafood was sold to the Japanese market, with salmon, crabs, and herring the primary commodities. Canada, the United Kingdom,

France, and South Korea were also good markets in 1989, but the value of their imports was small, compared to Japan's purchase of West Coast products. Consumers in the United States spent an estimated $28.3 billion for fishery goods in 1989, a 5 per cent increase from 1988.

The total included $19.1 billion in expenditures in food service establishments (*e.g.*, restaurants, carryouts, caterers); $9.0 billion in retail stores (for home consumption); and $181.7 million for industrial fish products. In producing and marketing a variety of fishery products for domestic and foreign markets, the commercial fishing industry contributed $17.2 billion in value-added dollars to the Gross National Product (GNP), an increase of 5 per cent compared to 1988. Consumption of fish and shellfish in the United States totaled 15.9 pounds of edible meat per person in 1989.

This total was up 0.7 pound from the 15.2 pounds consumed per capita in 1988. Per capita consumption of fresh and frozen products registered a total of 10.5 pounds, an increase of 0.3 pound from the 1988 level. Fresh and frozen finfish consumption was 7.1 pounds per capita in 1989. Fresh and frozen shellfish consumption amounted to 3.4 pounds per capita, with canned fishery products at 5.1 pounds per capita, up 0.4 pound over 1988. The per capita use of all fishery products (edible and non-edible) was 62.2 pounds (round weight), up 2.8 pounds compared with 1988.

Although most of the fish and shellfish consumed is from commercial production, a significant share is caught recreationally. In 1990, the National Marine Fisheries Service (NMFS) estimated that 17 million marine anglers harvested more than 600 million pounds of finfish. Although statistics are lacking, NMFS suggests that 200-300 million live pounds of molluscs and crustaceans was harvested by recreationalists. This catch represents 3-4 live pounds or about 1-1.5 edible pounds of domestic per capita consumption, outside the commercial figure of over 15 edible pounds per person. The source, handling, and distribution of the recreational catch are just beginning to draw attention. Indeed, because recreational anglers are not regulated as food producers/ manufacturers, there is concern about the use and distribution of this "recreational" resource.

Although it is difficult to give definite numbers for either the commercial or the recreational harvesting sector, some general observations can be made. Commercially, the trend is towards more efficient activity. Consequently, the number of participants in the commercial sector is decreasing. The commercial processing industry appears headed towards consolidation, with increased dependence on imported products and aquaculture. Recreational participation remains strong.

Consumption data, as suggested by both the Department of Agriculture and the Department of Commerce, indicate a continued, if not expanding, harvest of sport caught fish and shellfish. More than 20 per cent of all fresh and frozen seafood consumed in the United States may now be attributed to non-commercial harvest and distribution.

GROWING MODE OF PRODUCTION IN THE SEAFOOD INDUSTRY

Aquaculture is a rapidly growing mode of production in the seafood industry. Annual production of farmed fish and shellfish in the United States has grown 305 per cent since 1980. The greatest production is of catfish. Catfish production increased 31 per cent from 1986 to 1987. According to the Catfish Institute, farm-raised catfish increased from 5.7 million pounds in 1970 to 295 million pounds in 1988 and were expected to exceed 310 million pounds in 1989. Salmon production in the Pacific Northwest and Maine totaled 85 million pounds in 1987. In addition, other fish that are farmed include trout, redfish, sturgeon, hybrid striped bass, carp, and tilapia, as well as shellfish and crustaceans such as oysters and crawfish. Crawfish production acreage has increased 145 per cent to about 160,000 acres.

Overall U.S., aquaculture production of fish and shellfish increased from 203 million pounds in 1980 to some 750 million pounds in 1987. It is estimated that by the year 2000, that figure will reach 1.26 billion pounds. Large amounts of cultured fish and shellfish are also imported annually. Approximately one-half of the 500 million pounds of shrimp imported is cultured; 143 million pounds comes from China and Ecuador, neither of which regulates the use of chemotherapeutic agents in culture. More than 40 million pounds of salmon is also imported annually, often from countries similarly lacking tolerance levels for residues.

Of special interest are the use of chloramphenicol in shrimp culture and ampicillin in yellowtail culture. The Food and Drug Administration (FDA) has not examined imported seafood for drug residues, and there is no information regarding levels that might be ingested.

Aquaculture also produces fish used to stock recreational fishing areas. This procedure is under the control of government agencies that follow FDA regulations, use only approved drugs, and abide by legal withdrawal times.

CONSUMPTION TRENDS

Today's consumer is changing rapidly. Instead of single-income households, it is increasingly more common to have both man and woman working. The size of the family is decreasing. As many as one-fourth of all households are occupied by one person. This means more shoppers and diners, most with little time for home preparation.

Most adult men and women now work outside the home. In recent surveys, 7 out of 10 new home buyers noted that they will need two incomes to pay their respective mortgages. Nevertheless, the growth in two-income couples has generally created an increase in disposable income, but with little time to spend it. With as many as 50 per cent of new mothers working outside the home within the first year of childbirth, it is easy to see the revolutionary changes taking place among families. The working mother or single dweller does not have the time to prepare meals in the traditional sense. In recent Food Marketing Institute (FMI) surveys, more than 30 per cent of the husbands of women who work full-time did as much cooking, cleaning, and food shopping as their wives.

The population is aging. Going into the next century, the fastest growing groups will be those aged 45 to 54, along with those over age 85. By the year 2000, the proportion of Americans over age 65 will be the same throughout the country as the proportion in Florida today. An aging population means decreased discretionary spending and more demands for healthful and nutritious foods.

Minorities are growing in America. Within 10 years, one-quarter of all Americans will be either black, Hispanic, or Asian. The city of Los Angeles illustrates the trend. At present, Los Angeles is the largest Mexican city outside Mexico, the second largest Chinese city outside China, the second largest Japanese city outside Japan, and the largest Philippine city outside the Philippines.

The consumer demand for convenience, gourmet foods, ethnic items, and other services is increasingly evident in the food service and retail food industries. As the number of working women and single dwellers increases, the consumer base continues to change. With reduced leisure time, consumers who once spent two hours per day in the kitchen now spend less than a half hour. Convenience stores, fast-food restaurants, specialty food service outlets, and prepared items in the supermarket are food industry responses.

To illustrate the impact of less preparation time in the home, a quick review of consumer buying habits is in order. In 1973, almost 80 per cent of the food dollar was spent on home-prepared foods. In 1988, this number had fallen to 67 per cent. Many predict that the figure may be as low as 40 per cent by the year 2000. As with all foods, fish and shellfish preparation must be viewed in the manner in which consumers use the product in a contemporary environment. This does not mean that the consumer will be eating at home less but, rather, that less time will be devoted to food preparation. This trend towards "cocooning," in which the family spends more time around the home but utilises the time more prudently, is central to future consumer patterns. Consumers want more convenience and nutrition. Value-added products, ready-to-eat items, and microwave entrees are examples. Deli departments of the supermarket may soon become food service operations, competing with fast-food and takeout restaurants.

Seafood, like other foods, will be placed in a competitive consumer environment. Fish and shellfish must continue to taste good if they are expected to attract more consumers. Further, seafood must stay within the budget of the new consumer. If the industry can respond to the changing consumer base, the opportunity to expand per capita consumption appears good.

The amount of imported product is not yet recognised as a potential problem by the consumer, yet it is of significant concern to regulatory officials. Rising needs place increased pressure on government to protect consumers without the ability to monitor the harvest, processing, and distribution of the hundreds of species in question. Because of the potential of ever-increasing imports, the safety issue is becoming a matter of international concern. Although agencies routinely sample and require country-of-origin labeling, the consumer is unaware of the complexity of attempting to truly safeguard these foodstuffs.

CONTRIBUTED TO MAINTAINING OF MARINE FISHERIES

The marine fisheries have contributed to maintaining fish supplies by offsetting the decline in catches of the inland fisheries sector. Over the last two decades the share of marine fisheries in total national landings has gone up from 10.6 per cent in 1970 to 28.1 per cent in 1990.

Bangladesh has an extensive continental shelf extending virtually to the edge of the 200 mile Exclusive Economic Zone (EEZ). The total area of the shelf down to 200 meters is estimated at 67,000 km2 of which an estimated 37,000 km2 are of less than 50 meter depth. This area is exploited by the country's small scale fleet composed of some 17,000 boats of different sizes, 6,000 of which are reported to be motorized. About 90 per cent of the marine fish catches are taken by artisanal fishermen and the remaining 5 per cent by the industrial trawlers. However, in terms of value, the industrial sector is of great importance because landings of high-valued shrimp. But at the same time it is responsible for discarding about 30,000 tons of by catches (or trash fish) per annum.

The offshore industrial trawl fishery is relatively a new development, and the Government involvement in the marine fishery field is through the Bangladesh Fishery Development Corporation (BFDC) which operates a fleet of fish and shrimp trawlers. Except for the few trawlers owned by the BFDC most are privately owned, some being operated on partnership basis. Since 1974, there have been initiatives taken by private commercial operators for off-shore fishing, with special emphasis on shrimp trawling for export. To date some 70 licenses have been given by Government for trawlers of different sizes. Only 53 are currently in operation, as some are still under construction, while some of the older ones are laid off for major repairs. of the 53 trawlers in operation, 40 are engaged for shrimp trawling and others for fish. Overall the shrimp catches have dropped recently, possibly due to excess fishing effort. From a peak of some 5,000 in 1988, the catch dropped to 3,200 tons in 1990.

No attempt has been made to exploit the off-shore and deep sea pelagic stock of tuna, tuna-like species, shark, etc. and these stocks are reported to be exploited illegally by foreign vessels. Exploitation of these resources is constrained by the lack of knowledge of fish resources and of the means of exploitation, and absence of any reliable catch and effort data. The fishery resources of the Bay of Bengal are strongly influenced by highly changeable environmental conditions. Generally the period from September to March is characterized by fine weather and calm seas. In April, the south-west monsoon appears and the period April to August is marked by rough seas, caused by strong winds with heavy rain, resulting in considerably reduced fishing operations. The Bangladesh coast is often subjected to heavy cyclonic storms during May and June and again during October to December.

Majority of the shrimp species exploited are common to both artisanal and industrial fisheries. The industrial fishery harvests mostly the adult phase and the artisanal the preadult, post juveniles and the juveniles. Penaeus monodon is the most targeted species and fetches a very good price both in local and international markets. However, the highest contribution to total production (63 per cent) is made by Metspenaeus monoceros (brown shrimp). The finfish species presently exploited are mainly demersal, shallow water estuarine species and also some mid water species. Hilsa is the prominent fish in both inland and marine catches accounting for about 13 per cent of inland, 46 per cent of marine and 22 per cent of total fish production. Hilsa accounts for 78 per cent of the fish caught by mechanized gillnets which in turn account for 94 per cent of the small-scale marine fishery. A number of surveys have been conducted in the marine waters of the Bangladesh continental shelf but controversies remain about the extent of fish resources within the EEZ, particularly so far as the pelagic fishery resources are concerned. The 1979–80 survey assessed that the pelagic stock was between 60,000–120,000 tons. Surveys of demersal resources give different estimates of the demersal fishery biomass varying between 55,000 tons and 373,000 tons. However, the results of the three most recent demersal stock assessment surveys, carried out at the exploited phase of the stock, are quite close between 152,000–160,000 tons. A reliable assessment of the penaeid shrimp stock has yet to be made, there are large variations in the estimates ranging between 1,000–9,000 tons.

The World Bank Review indicated that marine fish production could increase from its present level of 230,000 tons to between 300,000 and 325,000 tons by 2010 at an annual growth rate of 1.3 to 1.6 per cent per annum. Such an increment could be realized through the improvements in fishing technologies for pelagic fisheries (tuna, tuna-like species, sharks) which could amount to about 25,000 tons per year; improving Hilsa stocks (which is both inland and marine) through rehabilitation and conservation which could and another 25,000 tons/year; harvesting fish now allegedly caught by foreign trawlers which is estimated around 15,000 tons per annum in the Bangladesh waters; and collecting the shrimp trawlers by-catches amounting some 30,000 tons/year.

FISHING COMMUNITY

Social Status

Fishing in Bangladesh was an occupation undertaken primarily by members of particular Hindu castes. Given the low social status associated with fishing, these communities traditionally occupied the lower rungs of the social hierarchy in rural communities. Even in Bangladesh Muslim society, where caste is not recognized, groups involved traditionally in fisheries are accorded a generally low social status. Significant changes have taken place in the social groups involved in fisheries with many Hindu fishermen migrating to India over the last 40 years, increased population pressure and the steady contraction of fisheries resources.

While a core of full-time professional caste fisherfolk are still to be found in most floodplain areas of the country, their traditional field of activity is increasingly opening up to a wider range of rural people who have become more and more involved in fisheries.

These newer entrants to fisheries may be attracted by the profitability of certain fisheries activities, at least on the seasonal basis, or they may be driven into fisheries through their exclusion from other labour markets or areas of income earning activity.

Fishermen Population and Group Formation

It is estimated that there are about 768,000 inland water fishermen who are often concurrently farmers, and 510,000 marine water fishermen, the majority of whom being traditional artisanal fishermen. They operate fishing in about 870 coastal villages and 45,000 rural communities in inland areas. Those who catch and appropriate fish may be grouped into four:

a. traditional caste fishermen,
b. non-traditional fishermen,
c. Jalmahal lease-holders, trawler or mechanized boats owners (who are not fishermen); and
d. the public who catch fish for subsistence purposes.

Fishing is normally carried out on individual basis, but in some instances fishermen form mutual-aid-groups out of kin. The capital is shared equally among the members of the group. In a larger scale fishing, formation of a group may require the inclusion of outsiders. The normal size of the group varies from seven to twelve members from the same kinship group.

Fishermen's Cooperatives

Fishermen's cooperatives were formed by the initiative of the government to promote welfare of fishermen. However, in general, these cooperatives are ridden by unscrupulous non-fishermen (*e.g.,* fish traders) and they find their own position and financial resources. In the absence of effective supportive structure within the cooperatives, fishermen members neither play a useful role nor get any benefits from the cooperatives. There are three types of fishermen's organizations in Bangladesh:

a. Bangladesh Jatio Matshyajibi Samabay Samity (BJMSS), established in 1960;
b. Bangladesh Jatio Matshyajibi Samity (BJMS), established in 1986;
c. Bangladesh Jatiotabadi Jele Dal (GJJD), established in 1993.

The last two societies were established at the national level under different government departments; they do not have access to societies at field level.

The BJMSS is the apex society registered with the Department of Cooperatives and had 88 central and 4,243 primary societies in 1992. The total individual membership of the primary societies numbered about 540,000.

POST-HARVEST TECHNOLOGY AND MARKETING

Fish Processing

The country's fish processing industry, mainly freezing plants. has been over-capitalized. From 16 plants of a total capacity of 135 tons/day in 1980, the number has increased to 93 plants with capacity of 650 tons/day, and their utilization is estimated at around 15–20 per cent of the capacity because of lack of raw material.

The main reason for the overgrowth of the industry the availability of favourable and liberal credit. The bulk of the fish produced in the country is consumed domestically without any processing and only an estimated 25,000 tons of shrimps, frog legs and fish are processed for export annually. recent study estimate that 22 units working at 60 per cent capacity would be adequate to handle what is currently produced and exported. On the other hand there is a grave shortage of ice for the fishing industry, in particular along the coastal belt and along market routes. Altogether there are 215 ice plants with a combined capacity of some 4,000 tons per day. The shortage is most felt during the peak fishing season as evidenced by the price variations from TK 20 to TK 300 a block. With a projected increase in supplies from recently established shrimp farms and in shrimp and fish exports, adequate supply of ice to preserve fish and shrimp from harvest to processing is extremely essential.

There has been very little attempt by the government to provide support services and infrastructure to the traditional fisheries utilizing thousands of small mechanized and non-mechanized craft, ranging from small canoes to large motorized craft, although they supply about 25 per cent of the country's fish supply.

These traditional fleets have no adequate bases, or means supplies. The traditional boats which resort to day fishing do not normally carry ice and those vessels with extended fishing days, carry ice but much below the required norms. Loss of fish due to bad handling and ill-equipped landing sites and markets has been estimated at 15 to 20 per cent.

The trawler industry is based at Chittagong, and the fishery harbour, constructed with apparent disregard to hydraulic engineering feasibility has silted up and the industrial fishing fleet has to face major operational hazards, with loss of fishing time and fish quality.

Fish Marketing

Fish is marketed in many forms in Bangladesh, ranging from fry/fingerlings for culture to live, fresh and dried/cured fish for domestic consumption and frozen products for export. Along with the aquaculture development in recent years, trade in juvenile fish and crustacea has significantly increased.

Fish marketing is carried out at four different stages, *i.e.*, primary villages, assembly points, secondary distribution points, and terminal cities. Fish marketing, which is largely in the hands of the private sector, is managed, financed, and controlled by a group of intermediaries known as aratdars (commission agents) and mahanjans (financiers or money lenders). Wholesale fish markets are mostly run by a few aratdars who greatly restrict the newcomers.

The aratdars provide advances to fish traders who in turn are required to bring fish to them for sale. The fish traders also provide advance to the fishermen who are to sell fish to them. Competition is not keen especially at the fish assembly stage. The assembling of fish from fishermen is perhaps the most profitable activity in the entire marketing chain, because the fishermen lack bargaining power. A study carried out by the Overseas Development Administration (1987) indicates that the village traders (primary assemblers) normally charges a commission of 30 per cent of the retail price.

There is evidence of widespread exploitation of fishermen and the extraction of rent by traders and leaseholders. It will, however, be difficult to reduce this exploitation unless competition is facilitated and fishermen are provided with adequate credit and marketing facilities and they develop a collective marketing system.

The fish marketing system is a traditional, complex and less competitive but relatively efficient system. however, it faces serious problems, including heavy losses and waste and poor fish quality. Yet, since fish demand generally exceeds supply, there is limited incentives for traders to improve the quality of the marketing system.

It is believed that in Bangladesh 10 per cent to 12 per cent of all fish caught spoils and becomes unsuitable for human consumption. This economic waste can be reduced by cold storage facilities, insulated and refrigerated transport systems, and adequate supplies of ice. Fish is caught in small quantities throughout the country. The fish marketing infrastructure is inadequate and needs improvements to increase marketing efficiency. The marketing system must also adjust to:

a. The expanding export demand for quality frozen seafood,
b. The expanding domestic demand for quality fresh and frozen fish,
c. The large seasonal fish catches in areas far from the main markets and fish landing centres, and
d. Expanding fish supplies in development project areas.

The private sector accounts for most fish marketing, but a very small fraction is handled by the Bangladesh Fisheries Development Corporation (BFDC). BFDC originally entered the fish marketing to develop a modern fish marketing infrastructure, increase competition and provide wholesale and retail market stability to protect both fishermen and consumers. In addition to marketing the fish caught by BFDC trawlers and from the BFDC managed Kaptai Lake (BFDC carries out stocking and harvests around 3,500 tons per annum), BFDC buys fish from Oxbow Lakes Projects and from government project areas.

MARINE BIODIVERSITY

The need to maintain healthy ecosystem function is well understood in theory but often poorly achieved in practice. The history of post-World War II fishery management illustrates the problem. Worldwide, fisheries and their management have contributed to marine biodiversity loss and have resulted in large losses in the economic productivity of ecosystems.

The current state of world fisheries is both a cause and an effect of the biodiversity problem. Fisheries are a fundamental cause of the biodiversity problem because of the general focus on single-species management, slow response to environmental signals, and the patterns of sequential harvesting, overcapacity and failure to control access. Fisheries bear the effect of biodiversity loss through declining biological and economic productivity, loss of aesthetic and recreational values, a diminished range of harvested species, regulations implemented to protect endangered species, and increasing costs of management designed to protect further biodiversity loss (National Research Council, 1995).

World fisheries developed rapidly after World War II under national policies to promote expanded economic opportunities and control of ocean territory. Access remained open to all, subsidies were provided for the construction of vessels, and seafood markets expanded. As a result, fisheries were transformed from being 60 per cent underexploited in the early 1950s to 60 per cent overexploited by the early 1990s. FAO reports that of the fish stocks accounting for the majority of the world's marine landings, 60 per cent are in urgent need of management. The degree of overfishing varies widely by geographic area, but where it exists, species abundance is reduced and ecological resilience is compromised. Ecosystem productivity in coastal areas is also suffering from habitat degradation and fragmentation, and these declines in biological productivity translate directly into economic losses.

Biological and economic stresses on marine ecosystems also affect marine fishery management at all levels. Increasing conflicts, complicated regulations, expanding information requirements and declining legitimacy of management all contribute to increasing fishery management costs. The 1990s have seen a number of publications describing fishery management as in crisis (McGoodwin 1990; Crean and Symes 1996; Hannesson 1996).

Problems in maintaining the biodiversity of marine ecosystems also affects equity and fairness, issues that are especially relevant for food security and employment. Globally, fish are 16 per cent of the animal protein content of human diets. Regionally, fish are a more important protein source for poor countries in Africa (21 per cent) and centrally planned Asian economies (22 per cent) than for rich countries in Western Europe (10 per cent) and North America (7 per cent). When biological productivity of marine ecosystems falls, small-scale artisanal fisheries in poor countries suffer loss of subsistence and markets, often without available substitutes.

IMPLEMENTING BIODIVERSITY PROTECTION FOR COMMON PROPERTY FISHERIES

New institutional approaches to biodiversity protection will be done, if at all, within the hierarchy of the existing institutional environment that includes international agreements, national policies, government agencies, decision bodies, regulations and enforcement.

International Agreements

At the top levels there are many international agreements and national laws that support or promote biodiversity protection in fisheries. These include the Rome Consensus on World Fisheries adopted by the FAO Ministerial Meeting in March 1995, the Code of Conduct On Responsible Fisheries, adopted by the Rome Conference of FAO in October 1995, and most recently and most specifically, the Kyoto Declaration adopted at the Conference on the Sustainable Contribution of Fisheries to Food Security. The Rome Consensus includes agreement on the need to eliminate overfishing, reduce fishing capacity, reduce bycatch and discards and strengthen governance. The Code of Conduct contains guidelines on fishery management and operations, aquaculture, coastal zone management, trade and research. The Kyoto Declaration, concerned with food security, includes agreements on the need to reduce fishing capacity, strengthen the scientific basis for multispecies and ecosystem management, reduce incidental catch and strengthen institutional coordination. These international agreements reflect the environment of globalization. In several cases they are leading to national efforts to define and implement the various actions.

National Policies

Moving down to the level of national policies, the performance of fishery management nationally is critically important to the performance of fisheries management globally, because the majority of world fishery production comes from the continental shelves. In addition, nations bring their histories and expectations about common property resource governance to international agreements which are the basis for the control of fishing practices and the protection of habitat quality. An example of a recent national law strengthening biodiversity protection is the U.S. Sustainable Fisheries Act of 1996, which contains much stronger conservation protection measures related to essential fish habitat and overfishing. These measures are still new and largely untested. The specifics of implementation are still being worked out.

Operating within the framework of national laws are national and regional resource agencies, decisionmaking bodies, regulations and enforcement. All of these comprise institutional implementation. As one moves down the nested hierarchical structure the difficulties of institutional redesign increase at each successive level because actions increase in specificity and in the number of unresolved implementation issues. It is in this institutional implementation environment, at the agency level and below, that the real challenge of institutional design for common pool resources lies.

Agencies

Worldwide, fishery agencies range in level of authority from international intergovernmental, to national governmental, regional and local. At all levels these agencies, as a result of the heavy use of marine ecosystems, are struggling to adjust to a new institutional environment that involves redefined authorities, expanded responsibilities, continued fragmentation between agencies, expanded information requirements, and the complexities of new management directions. To implement biodiveristy protection, these agencies must develop new mechanisms to coordinate ecosystem-level management across fragmented authorities, implement data collection on the status and health of previously unmonitored species, and find methods to finance their new and broader ecosystem responsibilities.

Decision Bodies

Decision bodies range in structure from authoritarian, in which decisions flow from the top down, to participatory, in which users contribute to decisionmaking as advisors or as co-managers. Regardless of structure, decision bodies in fisheries are coping with similar problems in the implementation of biodiversity protection. Such bodies tend to have developed a narrow and shifting management scope does not account for biodiversity. They also tend to be operating within a cumbersome management structure that makes flexibility and adaptation difficult. As a result, many of the decisionmaking bodies are under strain at a time when they must accommodate a broader range of responsibilities and interests. The scope of management is often narrowly defined for single species, and vacillates from goals of conservation to maximizing employment to protection of historical fishing shares. Interests within fishery management-gear groups, operations of different scales, fishing communities, scientists, environmental organizations-are increasingly polarized, with ensuing conflicts that are costly and often intractable. Questions are being raised about the appropriate level of decisionmaking authority, with many recommendations for devolving control to as low a level as possible. The complex decision structures designed to maintain procedural transparency and protect against special interest capture often work against decision flexibility to changing environmental conditions, and promote slow adaptation to changing institutional conditions. It is at this level that the implementation of theory into practice is most sorely tested.

Regulation

The depleted condition of many fisheries has led to new regulations concerning the protection of essential fish habitat, overfishing, and the ecosystem effects of fishing. In the United States, regulations derived from measures in the 1996 Sustainable Fisheries Act are for the most part new and untested, are layered onto old systems of regulations, and carry system-wide effects. New measures for biodiversity protection are fundamentally different from existing regulations in several ways.

They integrate commodity use and Non-commodity existence of fish species. They introduce requirements for protection that are based on boundaries that are ecological rather than political or geographic. They require scientific information on species that are not directly exploited. They require protection of a wider range of species about which there is significant scientific uncertainty about critical functions and thresholds. And, perhaps most importantly for institutional redesign, they explicitly require accounting for interactive effects within marine ecosystems and, by implication, require more regulatory flexibility at a time when the overexploited state of many fisheries has led to regulatory rigidity.

Enforcement

Many of the regulations designed to protect biodiversity are enforceable only at sea. Such measures include restrictions on fishing technology and the deployment of gear, the prohibition of fishing in protected areas, temporary area closures, bycatch limits, prohibited species restrictions, and requirements for full retention (zero discards). At-sea enforcement, while ideal for monitoring behaviour within these types of regulations, is expensive. Fishery authorities currently rely heavily on onshore enforcement of regulations, most of which control access, gear use, and landings. Given the extent to which fishery agency budgets are already stretched and the low probability of budget expansions, effective at-sea enforcement of biodiversity protection measures becomes a serious challenge. It is in the enforcement arena that it becomes most important to have users vested in the goals and objectives of management.

INDUSTRIAL MARINE FISHERIES

This summary of the history of marine fisheries in Southeast Asia traces the development of fisheries in the region from basically subsistence activities in the nineteenth century to large-scale industrial fishing in the latter part of the twentieth and early twenty-first centuries and examines the impact that this development has had on fish stocks and fishing communities in the region. The work is based substantially on the comprehensive review by Butcher although other sources have also been used as appropriate. Units of measurement have, where possible, been standardized to metric units to facilitate comparisons with current and other data, although Butcher's units for weights and measures of measurement are already quoted in metric units except where other units are referred to in a statute, law or other legal instrument. However, there are instances, particularly for the measurement of vessel tonnage, where such standardization has not been possible and these instances are highlighted and discussed as they occur.

The geographical area covered by this history includes all of the countries of Southeast Asia, excluding China to the north and Australia to the south. The time period covered is approximately from the mid-nineteenth century to the present day. Although inland and estuarine fisheries are also important in Southeast Asia, the seas of the region have been particularly important and have long provided people with a variety of fish, shrimps, squids, whales, pearl oysters, sea cucumbers and a multitude of other animals that they have collected and captured for medicine, oil, jewellery and most importantly food.

Prior to the early 1900s, the population of the whole of Southeast Asia was about 40 million with the majority of that population living in villages. The countries of the area were not yet urbanized to any great extent and therefore fishing (both marine and inland) was directed mainly at the need to supply food for these village communities. There was some limited trade (often between coastal and inland villages) and simple preservation and processing methods such as salting, drying and the manufacture of fish sauce which were developed and used to facilitate this trade. The supply of fresh, preserved or processed fish for large centralized markets, either domestic or international, was largely unknown. As a result of this subsistence fishing by coastal villages in the nineteenth century, the region's marine fish stocks were almost certainly lightly exploited and, although no detailed surveys were carried out until after 1945, several contemporary nineteenth century records comment on both the quality and the abundance of fish in many areas and on the simple fishing methods used in marine fisheries.

Restricted by the simple fishing gear and vessels, and with abundant coastal fish resources, most fishing was undertaken in near-shore waters. When the demand for fish and other marine animals rose or the supply fell (often as a result of natural causes, such as the monsoons), it was a relatively simple matter to expand operations into new coastal fishing grounds since there were very few barriers to such expansion. As fishing moved away from a subsistence-level activity and became increasingly oriented towards supplying more remote markets in the late 1800s and early 1900s, a spectacular growth in catches occurred. This expanding demand for marine products was created above all else by urbanization and the growth in population but also by the development of transport and marketing systems and changes in the techniques in preserving fish, shrimps and other marine life that prompted people to produce for the market.

This expansion of marine production accelerated after the Second World War as fleets were mechanized and fishing activities expanded to new areas, particularly those areas offshore that had previously been unfished or, at least, only lightly exploited. The expansion of marine landings in Southeast Asia has been so spectacular that two nations in the region (Thailand and Indonesia) are now among the world's top 10 fish producing nations and marine fisheries production accounts for more than 1 percent of GDP in almost all countries of the region.

With this expansion in landings through the twentieth century came an expansion in the areas fished by national fishing fleets, a decline in the abundance of many fish stocks and, accordingly, an increase in interest at the regional, national and international level, in administering, controlling and regulating fishing activities in the region. States, and since the late 1940s, international organizations have all influenced the scale and location of fishing in the region and, until recently, fishing often took place beyond the reach and grasp of state powers. The ability to expand landings by moving to new areas was curtailed in the latter part of the twentieth century not only by the lack of new fishing grounds to move to but also very much by this increasing interest of national Governments (and the legal obligations and powers conferred by the UN Convention on the Law of the Sea, UNCLOS) to control fishing activities, particularly fishing by other nations' fleets, within a nation's territorial waters.

Such recent attempts at regulating fisheries have not been wholly successful since they are built on a long history of a lack of government intervention and control in the marine fisheries sector. However, as Butcher points out, the era of opening of new "frontiers of fisheries has ended. The challenge now is to exploit the seas in a sustainable manner that preserves the diversity of marine life while providing the people of the region with a source of food long into the future". In examining how the various sectors of marine fisheries in Southeast Asia have developed since the nineteenth century and how they have contributed to the overall rapid expansion of landings, the individual components that comprise the region's marine fisheries sector will be examined. These components include: (a) pearling; (b) trawling; (c) purse seining; (d) shrimp trawling; (e) tuna longlining, poling and purse seining; (f) driftnetting; (g) trolling; and (h) other industrial fishing operations, including failed types of industrial fishing. In addition, the current state of each of these sectors and the resources upon which they rely will be examined so that an assessment can be made for each sector of the likelihood of being able to move from an expansionist mode of fishing (where landings were increased by exploiting new areas) to one of sustainability, based on existing resources.

Pearling

In the nineteenth century, prior to the development of the cultured pearl industry in Japan, mother of pearl shells and pearls were some of the most valuable marine animal products of the time, and the demand for these products had been increasing rapidly in Europe and North America. In Southeast Asia, there was a thriving pearl shell and pearl industry in the early 1800s, based in the islands surrounding the Sulu and Celebes Seas which sold or bartered its products to China. So great was this trade that providing sufficient labour was a problem and therefore slave raiders were employed to capture slaves in the islands and coastal villages of the Malay Peninsula and bring them to Manila and Jolo. At the height of the trade in the 1830s, as many as 68 000 people were employed in diving for pearl shell in the area, more than the number of fishers in the whole of Java and Madura. About 730 tonnes of pearl shell was exported from Jolo each year. The pearl shell and pearls that were not exported directly to China (either by Chinese traders or by British East India Company vessels) were sent to the market at Makassar for sale.

The pearling areas around the Sulu and Celebes Seas, the majority of which is now encompassed within the Philippines, is still by far the major producer of pearl shell and pearls (as opposed to pearls from the pearl culture industry) in the region with pearl shell production in 2002 being about 25000 tonnes or over 99 percent of total production in the region. This production is, however, about the same level as the 1960s although landings have fluctuated considerably during that time.

The early, rapid development of the pearl shell and pearl industry in the eastern islands area (which almost exclusively used Japanese and Filipino divers) quickly resulted in the more easily accessible stocks coming under pressure and, as pearl stocks in shallower waters were intensively harvested, it became necessary to find ways to harvest in deeper waters to maintain and increase supplies. In 1839 Augustus Siebe developed diving gear which was eagerly adapted by companies with sufficient capital to purchase the necessary equipment. The addition of diving equipment not only enabled divers to reach greater depths (up to approximately 54 m) than had previously been possible with free-diving, but also to increase the proportion of the year in which it was possible to collect oysters from three to nine months. The development also enabled stocks in deeper waters to be exploited and, by the late 1800s, areas in the Aru Islands, the Mergui Archipelago and the Sulu Archipelago were being fished. Faced with the high costs of the pearling operations, pearlers generally adopted the strategy of extracting as much as possible as quickly as possible and then moving on to another oyster bed. This strategy was adopted in all three of the main pearling areas.

The first area where diving equipment resulted in an increase in yields was in the Mergui Archipelago where, according to an official report by Rudmose Brown and Simpson in 1907, there was "no systematic pearling in the archipelago before 1891 when the attention of the Government was drawn to these banks by a Queensland pearler [presumably from the Torres Straits]." To manage this fishery, the Government of Burma immediately introduced a "block system" whereby the government auctioned the rights to collect pearl oysters in five "blocks" covering the richest pearling grounds. As a result of this system the mentality was adopted that it was imperative to extract as much pearl shell as possible during the year as the same company would not necessarily secure the same block the following year. Official figures indicate that yields rose dramatically from 26 tonnes in 1891/92 to 340 tonnes in 1894/95, leading to the emergence of Mergui as a boomtown. By 1900/01 the yield had fallen to 66 tonnes, and the government abolished the block system and replaced it by a system under which pearlers bought a licence for each pump they used. There was apparently no limit on the number of licences used.

During this period the accepted view was that there was little danger of overexploitation of the Mergui pearl beds as the beds were continually restocked by the offspring produced by oysters in deeper waters beyond the reach of the divers. It was argued that there was a "natural balance" between the number of divers and the quantity of shell available for, as yields fell, divers would leave and the stocks would recover. As a result it was concluded there was no need for any sort of regulation of the Mergui pearl fishery.

Pearling also expanded into the Aru Islands where Australian pearlers, using Japanese and Filipino divers, adopted diving gear in the 1870s. The richest grounds were those along the eastern side of the Islands which, after a few failed attempts at development, were developed by the Australian pearlers from

Torres Strait in 1905, under a three-year concession granted by the Netherlands Indies Government. These grounds had hardly been touched by divers and, initially, catches were about twice what they were in the Torres Straits. In addition, the operations were significantly more profitable because licence costs, labour and the cost of living were cheaper than in the Torres Straits fishery.

This Aru Island fishery quickly expanded and by 1906 there were 150 vessels working in the area, operating as a fleet with multiple diving vessels supplying a schooner which acted as a "floating station" where sorting, cleaning and packing was undertaken. In 1906, indigenous divers in the area (who, under the terms of the concession, had exclusive access to pearl shell in waters less than 5 fathoms) supplied only 13 percent of the total 950 tonnes that were taken in the area. The company operating the fleet was the world's largest producer of pearl shell and, in the same year, supplied 37 percent of mother-of-pearl shell imported into London.

However, within a few years, yields began to decline as the three year concession provided no long-term incentives for preservation of the stock and contemporary anecdotal evidence appears to indicate a reduction in catch rates per vessel of about 50 percent between 1905 and 1908. Although the concession was actually renewed several times, the company operating the fleet had withdrawn most or all of its vessels by 1916. However, this withdrawal did not end pearling activities in the Islands with the indigenous, shallow water fishery continuing and a few diving vessels operating well into the 1930s. These activities continue today from smaller, motorized vessels.

In the Sulu Archipelago, the introduction of diving technology came later and in a milder form than in other areas and, in 1914, there were 73 pearling vessels operating there, 40 of which were owned by a Japanese company and the rest by individuals of various ethnicities. Only two of the Japanese vessels had motor-driven pumps for diving. By 1930, there were only 24 vessels operating, five of which were equipped with engines and moto-driven air pumps. The Sulu Archipelago area appears to have exported about 300 tonnes of pearl shell in 1914 and, throughout its development, was distinguished by a greater reliance on indigenous fishers operating in shallow water for most of its production. Yields do not appear to have declined as precipitously as other areas where mechanized diving operations were more important, and Butcher suggests that this may have been a result of the more extensive nature of the pearl beds and the apparently less intense exploitation.

The introduction of mechanical diving technology in the late nineteenth century therefore led to a boom-and-bust development of the pearl shell beds of the region with pearl beds being sequentially depleted. Throughout this period of development, however, it is interesting to note that little concern was expressed for the long-term sustainability of the stocks since it was commonly believed that depleted adult stocks would be quickly replenished from untouched "deeper water" stocks. The decline in abundance of pearl shell helped drive an increasing demand for trochus shell and other gastropods as a substitute for pearl shell which was used in the manufacture of buttons. But, after the Second World War, it was the development of cheap plastic substitutes for button manufacture that resulted in the decline of the industry although it continues today as a small, artisanal fishery in the Philippines.

CROP GENETIC RESOURCES

Presenting crop genetic resources at the global scale is an enormous task; this chapter will only attempt to provide an overview of findings and major global agreements, and present the status and progress made in certain critical activities.

Crop genetic resources include farmer's varieties and landraces, elite and special material and crop varieties developed by plant breeders and other researchers, wild and weedy relatives of crop plants, and wild plants harvested for food. They are used as raw material for the production of new crop varieties that respond to the needs of farmers and consumers. They provide insurance to meet future challenges posed by changes in the environment, diseases, and marketing opportunities, among others. Many crop cultivars also have significant cultural value for their holders.

The links between the conservation of plant genetic resources for food and agriculture (PGRFA) and its sustainable use as seed and propagating material, highlighting the key role of plant breeding. The system is farmer-centred as all the elements are integrated by farmers working in agro-ecosystems. Each component is related to the others, with germplasm flow and feedback loops connecting them.

Crop genetic resources systems typically acquire genetic resources, conserve their viability and genetic integrity, characterize diversity by evaluating its agronomic value, document all these activities, and facilitate their use by providing access to samples of material and associated information.

MAIN FINDINGS AND CHALLENGES

The first *Report on the State of the World's Plant Genetic Resources for Food and Agriculture.* In 1996, the first *SoW-PGRFA* Report was presented to 150 countries at the Fourth International Technical Conference on Plant Genetic Resources held in Leipzig, Germany. Prepared under the auspices of FAO, it was welcomed as the first comprehensive evaluation in this domain. An extended version of the first *SoW-PGRFA Report* was published by FAO in 1997. Although now almost ten years old, the first *SoW-PGRFA Report* remains the most widely cited source of information and assessment.

The first *SoW Report* pointed out that a small number of cereal crops provide a large proportion of total food requirements. However, when food energy supplies were analysed on a subregional level, a greater number and more types of crops emerged as significant. A substantial share of energy intake was also provided by meat, which is ultimately derived from forage and rangeland plants. For the most part, these plants were poorly collected, documented and exploited.

The first *SoW Report* provided summary information on many *ex situ* collections of major staple food crops. Over 1 300 gene banks were identified, preserving more than 5.5 million accessions, with wheat and rice accounting for 50 percent of the total. Continued gene flow between crops and their wild relatives in centres of origin underlined their importance as sources of new variability. Secondary centres of diversity were also very important. Various *in situ* approaches were identified: conservation of crop wild relatives and wild food plants; managed ecosystems such as rangelands; and conservation of traditional crop varieties on-farm and in home gardens. The report documented the facilitation of the use of PGRFA by evaluating agronomic characteristics, pre-breeding, crop improvement and seed supply programmes.

Fifty-nine countries reported national committees on PGRFA but some lacked a national programme of any type. Almost 80 percent of these countries referred to lack of training as a serious constraint. The first *SoW-PGRFA Report* also documented the status of regional and international collaboration in PGRFA, access to genetic resources and benefit-sharing. A technical annex reported on the state of the art of methods to analyse and assess genetic diversity and vulnerability, *in situ* conservation and methods for utilization of PGRFA through plant breeding.

Challenges

Conserving and using PGRFA effectively requires policies, knowledge and action. Crop genetic resources are at the intersection of agriculture, food security, environment and trade. A series of relevant global agreements affects their management, including the World Food Summit Plan of Action, the CBD, and global agreements on intellectual property rights. The exchange of crop genetic resources, from which everyone will eventually benefit, has been a reality since the beginning of agriculture. Our growing world population will only be fed if the widest possible range of genetic diversity can be draw upon. As a result, countries and regions are interdependent. The world greatly depends on crops that originated elsewhere for food and agriculture. No country can hope to do well with its own resources alone. Finally, it is a fact that PGRFA were developed and are maintained by humans - by farmers within their farming systems and by plant breeders and researchers. Unless diversity is conserved in gene banks, it is often lost when farming systems die or when scientific programmes come to an end. At the scientific and technical levels, challenges are posed by genetic erosion, genetic vulnerability and utilization. Value in crop genetic resources lies at the intra-specific level, in the diversity *within* a crop's gene pool. The first *SoW-PGRFA Report* defined genetic erosion as "the loss of genetic diversity, including the loss of individual genes, and the loss of particular combinations of genes (*i.e.*, of gene-complexes) such as those manifested in locally adapted landraces". There is still no scientific consensus on how to measure genetic diversity in a crop's gene pool. Similarly, there is no consensus on the optimal balance of *in situ* and *ex situ* conservation methods to combat genetic erosion. The first *SoW-PGRFA Report* identified measures to strengthen *ex situ* conservation, such as "the development of low-cost conservation technologies, and in particular, technologies for non-orthodox seeded and vegetatively propagated plants including *in vitro* methods and cryopreservation".

Genetic vulnerability has been described as "the condition that results when a widely planted crop is uniformly susceptible to a pest, pathogen or environmental hazard as a result of its genetic constitution, thereby creating a potential for widespread crop losses" (National Academy of Sciences, 1972, cited in FAO, 1996a). Genetic vulnerability pertains to the level of the crop genetic diversity actually being used. One of its main causes is the widespread replacement, since the 1950s, of genetically diverse traditional varieties by varieties with more homogenous genetic make-up. The first *SoW-PGRFA Report* noted that "there is no comprehensive or coordinated system for monitoring uniformity in agricultural species, and methodological tools which might help assess related genetic vulnerability have not been adequately developed". Over the past decade, farmers' reasons for maintaining or changing the varieties they grow have only just begun to be understood. Conservation of the diversity of crop genetic resources is not enough: their potential uses and values need to be understood by characterizing, evaluating and documenting them. Methods still need to be developed to improve and facilitate productive utilization. The first *SoW-PGRFA Report* noted that "a variety of plant breeding and biotechnological techniques, which often differ in technical complexity and cost, may be used in crop improvement... Biotechnological methods are now increasingly available to facilitate wide crosses thus allowing the introduction of the desired genes". It also noted that "while a number of countries have initiated crop improvement programmes based on new biotechnologies, not all countries have the capacity to use such technologies". There are biotechnologies relevant for every objective of a crop genetic resources system.

Another set of challenges is posed for taking action: Is there sufficient capacity for this? How can capacity be built where it is currently inadequate? How can the necessary cooperation be organized among countries and among disciplines, particularly in order to link conservation and use of crop genetic resources? How can the resources needed to address these issues be mobilized?

5

Aquatic Ecosystem

AQUARIUM FISH DISEASES

There are literally hundreds of afflictions that can effect the health of your fish. The most common maladies seen in home aquaria are usually either bacterial or parasitic in origin. Fungal infections are also sometimes seen, and occasionally physical ailments.

Luckily, most fish ailments are easily diagnosed and can be treated with success. The most common of these afflictions are included here. How to prevent fish disease has steps you can take to reduce the possibility of disease and help to keep disease from spreading if it should occure. A table of contents is provided along with a diagnostic chart with links to appropriate medications.

Understanding how an aquarium and its filtration work to support aquatic life is vital in preventing fish ailments. The basics of life support are the same whether you have a freshwater aquarium, saltwater aquarium, or a mini reef.

FISH DISEASE

Fish ailments can be separated into 4 general types including bacterial infections, fungal infections, parasitic or protozoan infections, and physical ailments and wounds.

1. *Bacterial Diseases:* Bacterial diseases are usually characterized by red streaks or spots and/or swelling of the abdomen or eye. These are best treated by antibiotics such as penicillin, amoxicillin, or erythromycin.
2. *Fungal Diseases*: Comon fungal infections often look like gray or white fluffy patches. are
3. *Parasitic Diseases:* The most common parasitic disease called "Ich" can be treated most effectively with copper or malachite green in the right dosage. Most treatments will have copper as an ingredient. Many water treatments like "Aquari-Sol" will also contain copper as an ingredient. If the treatment you use is an anti-biotic or copper based, remember to remove all carbon from the filtration system.
4. *Physical Ailments:* Physical Ailments are often the result of the

environment. Poor quality water conditions can lead to fish gasping, not eating, jumping out of the tank, and more. Tank mate problems can result in nipped fins and bite wounds.

HOW TO PREVENT FISH DISEASES

Some steps can be taken to reduce the possibility of your fish getting a disease.

Following these precautions can also help keep fish diseases from spreading if they do occur.

- Buy only good-quality, compatible fish.
- Quarantine new fish before adding them to the aquarium. (A hospital tank can be used for this).
- Avoid stressing the fish with rough handling, sudden changes in conditions, or "bully" tankmates.
- Don't overfeed your fish.
- Remove sick fish to a hospital tank for treatment.
- Disinfect nets used to move sick fish.
- Don't transfer water from the quarantine tank to the main aquarium.
- Don't let any metal come in contact with the aquarium water.
- *Anti-biotics:* When using any anti-biotic, make sure the biological filtration in your aquarium is not destroyed. You want to be certain the treatment does not kill the nitrifying bacteria in your system at the same time it attacks harmful bacteria on your fish. Although most of the treatments available at the store state that they will not harm your biological filter, sometimes they wil. It is best to either monitor your ammonia and nitrite levels, or use an ammonia remover such as "AmQuel" to be sure your levels of ammonia don't become a problem.
- *Copper Treatments:* When using any medication which has copper as an ingredient, be aware that most plants will not do as well. Invertebrates, such as snails, can also be killed if the amount of copper is sufficient. Indeed, most snail removers are copper based.

DIAGNOSTIC CHART

If you notice something is wrong with your fish, a proper diagnoses is usually all you need to worry about. In most cases, a proprietary treatment (fish medications) purchased at a pet store will work very well.

Steps to diagnosing and treating your fish:

- *Symptom*: Match the symptoms your fish is showing with those listed under the 'Symptom' column.
- *Possible Cause*: Learn about each disease by clicking on the link under 'Possible Cause'.

- *Medication:* Find the treatment for the disease under the 'Medication' column.
- *Product Link*: Use the 'Compare Prices' column to compare similar products and merchant's prices.

RED PEST

Symptoms: Bloody streaks on fins or body.

Red Pest is called such because of bloody streaks that appear on the body, fins and/or tail. These streaks could proceed to ulcerations and possibly lead to fin and tail rot with, in severe cases, the tail and/or fins falling off. As the disease is internal, external treatments are usually not effective, except in very slight cases. In slight cases, treat the aquarium with a disinfectant and clean the aquarium as best as possible. Do not feed a lot while the aquarium is being treated. To disinfect, use acriflavine (trypaflavine) or monacrin (monoaminoacridine) using a 0.2 per cent solution at the rate of 1 ml per litre. Both disinfectants will colour the water, but the colour disappears as the disinfectants dissipates.

If the fish do not appear to respond favourably, discontinue disinfections. Then add an antibiotic to the food. With flake food, use about 1 per cent of antibiotic and carefully mix it in. If you keep the fish hungry they should eagerly eat the mixture before the antibiotic dissipates. Antibiotics usually come in 250 mg capsules.

If added to 25 grams of flake food, one capsule should be enough to treat dozens of fish. A good antibiotic is chloromycetin (chloramphenicol). Or use tetracycline. If you feed your fish frozen foods or chopped foods, try to use the same ratio with mixing. As a last resort add at most 10 mg per litre of water.

MOUTH FUNGUS

Symptoms: White cottony patches around the mouth.

Mouth Fungus is so called because it looks like a fungus attack of the mouth. It is actually caused from the bacterium Chondrococcus columnaris. It shows up first as a gray or white line around the lips and later as short tufts sprouting from the mouth like fungus. The toxins produced and the inability to eat will be fatal unless treated at an early stage. This bacteria is often accompanied by a second infection of an Aeromonas bacteria. Penicillin at 10,000 units per litre is a very effective treatment. Treat with a second dose in two days. Or use chloromycetin, 10 to 20 mg per litre, with a second dose in two days. Other antibiotics can also be effective. Kanacyn (kanamycin) will treat both bacteria at once. Maracyn (erythromycin) is effective against C. columnaris, and using Maracyn 2 (minocycline) in conjuntion with it will treat the Aeromonas bacteria as well.

TUBERCULOS–MYCOBACTERIOSIS

Syn: fish tuberculosis, piscine tuberculosis, acid-fast disease, granuloma disease.

Symptoms: Emaciation, hollow belly, possibly sores.

Tuberculosis is caused by the bacterium Mycobacterium piscium. Fish infected with tuberculosis may become lethargic, hollow bellied, pale, show skin ulcers and frayed fins, have fin and scale loss, and loss of appetite. Yellowish or darker nodules may appear on the eyes or body and may deform the fish.

The main causes for this disease appears to be over crowding in unkempt conditions; *i.e.,* poor water quality. All fish species could be susceptible though some are more susceptible than others. Those most susceptible are the labyrinth air breathers like the Gouramis, Bettas, and Paradise Fish. Others include Neon Tetras, Discus, and the Ram Cichlid.

- There is no absolute treatment. However the most effective treatment known for this disease is to treat with Kanamycin and Vitamin B-6 for 30 days. Kanamycin can be purchased at your local fish store. Liquid baby vitamins work well as s Vitamin B-6 source. They are available at your local pharmacy. Add one drop per every 5 gallons of aquarium water during treatment.
- If the treatment is ineffective, the best thing to do is destroy the infected fish.
- If either unkempt conditions or over crowding are the suspected cause, correct the condition.

It is possible for humans to contract this disease so we recommend using caution when dealing with it. Humans are very rarely are at risk from aquariums though. It is more common to contract this disease from public swimming areas or as a food contaminant.

DROPSY

Symptoms: Bloating of the body, protruding scales.

Dropsy is caused from a bacterial infection of the kidneys, causing fluid accumulation or renal failure. The fluids in the body build up and cause the fish to bloat up and the scales to protrude. It appears to only cause trouble in weakened fish and possibly from unkempt aquarium conditions. An effective treatment is to add an antibiotic to the food. With flake food, use about 1 per cent of antibiotic and carefully mix it in.

If you keep the fish hungry they should eagerly eat the mixture before the antibiotic dissipates. Antibiotics usually come in 250 mg capsules. If added to 25 grams of flake food, one capsule should be enough to treat dozens of fish. A good antibiotic is chloromycetin (chloramphenicol). Or use tetracycline. If you feed your fish frozen foods or chopped foods, try to use the same ratio with mixing. As a last resort add at most 10 mg per litre of water. Also, if unkempt conditions are the suspected cause, correct it.

TAIL ROT AND FIN ROT

Symptoms: Disintegrating fins that may be reduced to stumps, exposed fin rays, blood on edges of fins, reddened areas at base of fins, skin ulcers with gray or red margins, cloudy eyes. Tail and fin rot appears to be a bacterial infection of the tail and/or fins and may be caused by generally poor conditions, bully, or fin nipping tankmates.

If aquarium conditions are not good an infection can be caused from a simple injury to the fins/tail. Tuberculosis can lead to tail and fin rot. Basically, the tail and/or fins become frayed or lose colour. Over time the affected area slowly breaks down. First, attempt to ascertain the cause. Then treat accordingly. Also, treat the water or fish with antibiotics.

If added to the water, use 20 - 30 mg per litre. If the fish is to be treated add an antibiotic to the food. With flake food, use about 1 per cent of antibiotic and carefully mix it in. If you keep the fish hungry they should eagerly eat the mixture before the antibiotic dissipates. Antibiotics usually come in 250 mg capsules.

If added to 25 grams of flake food, one capsule should be enough to treat dozens of fish. A good antibiotic is chloromycetin (chloramphenicol) or tetracycline. If you feed your fish frozen foods or chopped foods, try to use the same ratio with mixing. As a last resort add at most 10 mg per litre of water. Also, if unkempt conditions are the suspected cause, correct it.

AQUATIC ANIMAL

When fish are living under good conditions with favourable water quality, nutritious food, and optimal temperatures, disease generally occurs only when new bacteria, viruses, or parasites are introduced into the fish's habitat. When fish are living under less than optimal conditions, their resistance to disease is reduced, and they may be infected by common diseases that are often present in the environment at low levels. For this reason, it is very important to make sure that we do all that we can to ensure a good environment for our aquatic animals and protect them from the introduction of exotic diseases.

GOOD ENVIRONMENT

For private ponds that are managed for recreational fishing or ornamental fish, the most important environmental considerations are these:

- Water quality
  - *Oxygen*: A problem when there are excessive nutrients, dense growths of plants or algae, when herbicides have been used, when fish are stocked and fed at high levels, and when the weather is hot and cloudy.
  - *pH*: Some kinds of soil chemistry can make your water very acidic or very alkaline.

  - *Alkalinity*: This is a measure of minerals in the pond that help keep the pH at the right level.
- Food availability
  - Proper stocking, liming, and fertilization ensure that there is adequate natural food.
  - Supplemental feeding of quality fish feeds helps too.
- Pond balance
  - Proper stocking not only ensures good growth but also may include fish that control aquatic plants and snails that carry fish parasites.

For commercial aquaculture ponds, tanks, and raceways, the most important environmental considerations are these:

- Water quality
  - Dissolved gasses including oxygen and carbon dioxide
  - Waste products like ammonia and nitrite
  - Alkalinity for pH control
  - Calcium and magnesium (hardness) and salt levels
  - Temperature appropriate for the species
- Food quality
  - Appropriate protein and fat levels
  - Quality ingredients
  - Correct pellet size and form
- Stocking
  - The right species
  - The right stocking density

PREVENTING EXOTIC DISEASES

For private fishing ponds, the primary consideration is the health of the fish that you put in the pond. For sportfish, this means that the fish should come from a reputable farm and that the fish should look healthy and free of disease when you purchase and stock them. It may even be worthwhile to ask the supplier about their fish health inspection programme. Many sportfish farms have very rigorous disease prevention programme, and they will be happy to share that information with you. You should never stock fish that you have caught in the wild. The wild fish may carry diseases that you do not want in your pond.

For ornamental fish ponds, the most important consideration is to buy healthy fish that come with some assurances about their disease status. Of special concern are koi. There are some very serious exotic koi diseases that you must avoid. If you buy healthy–looking fish produced in the United States by a reputable breeder, the chance of having problems with koi ulcer disease or koi herpes virus (KHV) are fairly low; however, new ornamental fish should always be quarantined for several weeks (preferably at water temperatures in the 70s) before they are put in your pond.

It would also be a good idea to ask the breeder about their koi herpes virus prevention programme. Good koi breeders will be happy to tell you about their disease prevention measures.

For commercial aquaculture, preventing the introduction of exotic diseases is of critical importance. All farms that produce aquatic animals should have plans to prevent the introduction of exotic diseases. The main routes of introduction are (in order of importance) infected animals, contaminated water, contaminated equipment, and wild animals that visit the farm facility.

To avoid disease problems, farms must produce their own stock or buy stock only from sources with good fish disease testing programmes and good disease prevention plans. If possible, the farm should use water supplies that do not harbor exotic diseases (wells and springs are very safe; run-off water is good; surface water from lakes and rivers may carry significant risks). Equipment returning to the farm from other places should be cleaned and disinfected before entering. Visits to the farm by wild animals that may carry diseases must be discouraged.

RECOGNIZING FISH HEALTH PROBLEMS

It can be difficult to monitor aquatic animals for health problems, especially when the animals live in large ponds or in water that is not crystal clear.

The list below describes some of the clues that may suggest that your aquatic animals are sick:

- A reduction in feed consumption not clearly related to low oxygen or temperature changes
- Any change in behaviour including:
  - Fish congregating at the surface or at the water's edge
  - Fish congregating around fresh water inflows or aerators
  - Fish swimming slowly or in circles
  - Fish "flashing" (rubbing against the bottom or other hard surfaces)
- Damage to skin and fins including:
  - Fin erosion
  - Skin ulcers
  - Skin discoloration (darker, lighter, reddening)
  - White or tan fuzzy patches
  - White spots
  - Bumps on the skin
  - Open sores
  - Cloudy or swollen eyes
- Dead and dying fish (the obvious one!)

SICKNESS OF AQUATIC ANIMALS

The first step whenever any aquatic animal health problem is suspected is to look at the water quality. Commercial fish producers usually have some water quality testing equipment. All producers should be able to test for oxygen levels, ammonia, pH, and nitrite and should recognize optimal and acceptable levels of these chemicals.

Sportfish farm pond owners and ornamental fish owners will often need to seek out professional assistance with their water quality testing. When water quality problems are ruled out, the next step is to consider parasites and other diseases.

Many aquatic animal diseases produce symptoms that are very similar. A correct diagnosis usually requires an aquatic animal health laboratory with the expertise to recognize fish parasites by microscopy and the ability to culture and identify bacteria, fungi, and viruses. When a disease problem is suspected, the first thing to do is to contact a fish health professional or fish disease diagnostic laboratory for assistance.

FISH DISEASE DIAGNOSTIC LABORATORY

Finding help with a fish health problem can often be quitc challenging. Assistance might come from state or university fish health laboratories, veterinary schools, extension programmes, wildlife agencies, or veterinarians who include aquatic animals in their practice.

For commercial fish farmers, the best way to find fish health assistance is to consult your state aquaculture coordinator, your state aquaculture extension staff, your state veterinarian, other local fish and shellfish producers, or the American Veterinary Medical Association at www.aquavetmed.info.

For private sportfish ponds, you may find help at the same places as the fish farmers, but it is also likely that your state wildlife agency will be able to guide you. Ornamental fish owners may find assistance at the same sources as fish farmers, but they may want access to services like X rays, ultrasound, blood chemistry, and surgery that are only available through veterinarians.

Once you have identified a fish health professional, the next step is to call that professional and ask what kinds of information and samples will be needed.

The information needed will often include answers to the following questions:

- How big is the pond/tank?
- What species of animals are dying?
- How many fish are dying?
- How long have they been dying?
- Are there sick fish swimming around right now?
- How are those sick fish behaving?
- What do they look like (sores, spots, and the like)?
- Are they eating?

- Have you used any treatments?
- Have you treated weeds or algae recently?
- Have there been any recent changes in the colour or clarity of the water?
- What has the weather been like (temperatures, wind, and rain)?
- When did you last stock new fish?
- When did you last harvest fish?

In some cases, help may come to your farm or pond, but often you will be asked to send samples to a laboratory. These samples may include water, live fish, fish on ice, or other materials needed for testing. Only the laboratory can provide you with detailed directions. It is very important to always call the lab prior to sending any samples for testing.

TREATING FISH DISEASES

Drugs and chemicals to treat diseases in aquatic animals are highly regulated by the FDA and EPA. There are very few legal drugs, especially for food animals, and very strict limitations on their use.

All drugs and chemicals must be used exactly according to their labels or under direct veterinary supervision. Any violation is punishable by state and federal law, and any product treated with an illegal drug or chemical is "adulterated" and would be ordered destroyed.

When treating an aquatic animal disease:

- Be sure that the disease has been correctly diagnosed (often this requires assistance from a fish health professional).
- Be sure that the chemical or drug to be used is appropriate for the disease, the fish species, and the water quality (including the temperature).
- Be sure that the chemical or drug is legal for the use.
- Be sure that the chemical or drug is used exactly according to the label or according to guidance from your veterinarian.
- For food animals, be sure that the required withdrawal period (time elapsed between the last treatment and sale of the animals as food) is followed.

For commercial fish production, there are also several legal vaccines that can be used to prevent fish diseases. As with drugs, these vaccines must be used according to their labels.

FISHERIES HISTORICALLY

Fisheries have been important parts of human life and food production throughout history. Fish acts as one main food in people's everyday life as well as food security of the nation. Fisheries have become a part of human cultures and mythologies, providing a community identity and a subject for artists throughout the ages. Partially, this is because fisheries are irretrievably wrapped up in humanity's perpetual fascination with the sea, and partially, because they have been a major source of food and income for many communities throughout the ages.

One of the world's longest lasting trade histories is the trade of dry cod from the Lofoten area to the southern parts of Europe, Italy, Spain and Portugal. The trade in cod started during the viking period or before, has been going on for more than 1000 years and is still important.

In India, the Pandyas, a classical Dravidian Tamil kingdom, were known for the pearl fishery as early as the 1st century BC. Their seaport Tuticorin was known for deep sea pearl fishing. The paravas, a Tamil caste centred in Tuticorin, developed a rich community because of their pearl trade, navigation knowledge and fisheries.

FISHERIES IN THE PRESENT DAY

Today, fisheries are estimated to provide 16 per cent of the world population's protein, and that figure is considerably elevated in some developing nations and in regions that depend heavily on the sea. Fisheries are a huge global business and provide income for millions of people. Fisheries have been and continue to be culturally important for many communities as well.

According to the Food and Agriculture Organization of the United Nations, total world capture fisheries production in 2000 was 86 million tons. The top producing countries were, in order, the People's Republic of China (excluding Hong Kong and Taiwan), Peru, Japan, the United States, Chile, Indonesia, Russia, India, Thailand, Norway and Iceland. Those countries accounted for more than half of the world's production; China alone accounted for a third of the world's production. Of that production, over 90 per cent was marine and less than 10 per cent was inland. There are large and important fisheries worldwide for various species of fish, mollusks and crustaceans. However, a very small number of species support the majority of the world's fisheries. Some of these species are herring, cod, anchovy, tuna, flounder, mullet, squid, shrimp, salmon, crab, lobster, oyster and scallops. All except these last four provided a worldwide catch of well over a million tonnes in 1999, with herring and sardines together providing a catch of over 22 million metric tons in 1999. Many other species as well are fished in smaller numbers, both locally and globally.

Methods

Fishing methods vary according to the region, the species being fished for, and the amount of income and technology available to the fisher. A fishery can consist of one man with a small boat hand-casting nets, to a huge fleet of trawlers processing tons of fish per day. Some common commercial techniques today are trawling, seining, driftnetting, handlining, longlining, gillnetting, and diving.

Trawling

Trawling is a method of fishing that involves actively pulling a fishing net through the water behind one or more boats.

Structure of a Trawl Net

When two boats are used (pair trawling), the horizontal spread of the net is provided by the boats, with one warp attached to each boat. However, single-boat trawling is more common. Here, the horizontal spread of the net is provided by trawl doors (also known as "otter boards"). Trawl doors are available in various sizes and shapes and may be specialized to keep in contact with the sea bottom or to remain elevated in the water. In all cases, doors essentially act as wings, using a hydrodynamic shape to provide horizontal spread. As with all wings, the towing vessel must go at a certain speed for the doors to remain standing and functional. This speed varies, but is generally in the range of 2.5-4.0 knots. Nets for trawling in surface waters and for trawling in deep water and over the bottom. Note the "tangles" with all of the marine life caught up in them.

The vertical opening of a trawl net is created using flotation on the upper edge ("floatline") and weight on the lower edge ("footrope") of the net mouth. The configuration of the footrope varies based on the expected bottom shape. The more uneven the bottom, the more robust the footrope configuration must be to prevent net damage.

Environmental Impacts of Trawling

Although trawling today is heavily regulated in some nations, it remains the target of many protests by environmentalists. Environmental concerns related to trawling refer to two areas: a perceived lack of selectivity and the physical damage which the trawl does to the seabed.

Selectivity

Reports of the lack of selectivity of trawling have been present since it started (about the 1600's) and it became used more widely (about 1900). Trawl nets may be non-selective, sweeping up both marketable and undesirable fish and fish of both legal and illegal size. Any part of the catch which cannot be used is considered as by-catch. Size selectivity is controlled by the mesh size of the "cod-end" - the part of the trawl where fish are retained. Fishermen complain that a mesh size which allows undersized fish to escape also allows a proportion of legal-landing sized fish to escape as well. There are a number of "fixes", such as tying a rope around the "cod-end" to prevent the mesh from opening fully, which have been developed to work around technical regulation of size selectivity. One problem is when the mesh gets pulled into narrow diamond shapes (rhombuses) instead of squares.

The capture of undesirable species is a recognized problem with all fishing methods and unites environmentalists, who do not want to see fish killed needlessly and fishermen, who do not want to waste their time sorting unsellable fish from their catch. A number of methods to minimize this have been developed for use in trawling. Bycatch reduction grills or square mesh panels of net can be fitted to parts of the trawl, allowing certain species to escape while retaining others. Trawling for shrimps has specifically been cited as having high levels of bycatch in various parts of the world.

Ecological Damage

Because bottom trawling involves towing heavy fishing gear over the seabed at a speed of several knots, it is destructive to the ocean bottom. The primary sources of impact are the doors, which can weigh several tonnes and create furrows when dragged along the bottom, and the footrope configuration, which usually remains in contact with the bottom across the entire lower edge of the net. Depending on the configuration, the footrope may turn over large rocks or boulders, disturb or damage sessile organisms or rework bottom sediments. Published research has shown that benthic trawling destroys the cold-water coral *Lophelia pertusa*, an important habitat for many deep-sea organisms. The primary focus of dispute over the impact of trawl gear is on the magnitude and duration of these impacts. Opponents of trawl gear argue that the impact of trawl nets is widespread, intense and long-lasting. Defenders of trawl gear maintain that impact is mostly limited and of low intensity compared to natural events.

Pelagic trawling is a much "cleaner" method of fishing, in that the catch usually consists of just one species. However, some cases of it have been attacked for depleting resources which are important sources of food for certain sea birds. An instance of this was the RSPB linking a population crash of sea birds in the North Sea to pelagic fishing for sand eels, which are food for many seabird species. This led to political pressure for the closure of this fishery; the seabird populations subsequently improved. However, a second population crash of seabirds which occurred with no effect on the stocks of sand eel, cast doubt on this link.

Other uses of the Word "Trawl"

The noun "trawl" has many possibly confusing meanings in commercial fisheries. For example, two or more lobster pots that are fished together may be referred to as a trawl. In some older usages "trawling" meant "long-line fishing"; that usage occurs in Rudyard Kipling's book *Captains Courageous*. (This use is perhaps confused with trolling, where a baited line is trailed behind a boat) The word "trawling" has come to be used in a number of non-fishing contexts, usually meaning indiscriminate collection with the intent of picking out the useful bits. For instance, in law enforcement it may refer to collecting large records of telephone calls hoping to find calls made by suspects. The word "trawling" occurs frequently in general literature and is used to mean searching through literature for information more often than it means catching fish.

Hand-line Fishing

Handlining is one of the oldest forms of fishing and is still common. The method consists of a single fishing line with a weight and one or more lure-like hooks are attached. The line is jigged or moved up and down in a series of short movements, most often close to the sea floor. The motion attracts the fish, which are normally caught while trying to eat the lure but also as they move close to jigged the lure. The line is then hauled onboard and the fish removed. Handlining are most often used to catch groundfish and squid but also other species are sometimes caught, including pelagic species.

Long-line fishing

Long-line fishing is a commercial fishing technique that uses hundreds or even thousands of baited hooks hanging from a single line. Swordfish, tuna and Patagonian toothfish are commonly caught by this method. It is also practiced on a smaller scale in New Zealand, where a twenty five hook maximum is prescribed by law. Longlines can be set to hang near the surface, for instance in tuna fisheries, or on the sea bottom, such as in the Patagonian toothfish fishery. Long-line fishing is controversial because the lines can lead to significant bycatch, often of endangered species such as sea turtles, petrels and albatrosses. Methods to mitigate such incidental mortality have been developed and successfully implemented in some fisheries.

These include the use of weights to ensure the lines sink quickly, the deployment of streamer lines to scare birds away from the baited hooks as they are deployed, setting lines only at night with ship lighting kept low (to avoid attracting birds), limiting fishing seasons to the southern winter (when most seabirds are not feeding young), and not discharging offal while setting lines.

GILLNET

Gillnetting is a common fishing method used by commercial fishermen of all the oceans. Because gillnets can be so effective their use is closely monitored and regulated by fisheries and enforcement agencies, such as the National Marine Fisheries Service in USA. Mesh size, twine strength, as well as net length and depth are all closely regulated to reduce bycatch of non-target species. Most salmon fisheries in particular have an extremely low incidence of catching non-target species.

History

Gillnetting began with First Nations fishermen using canoes and cedar fibre nets. They would attach stones to the bottom of the nets as weights, and pieces of wood to the top, to use as floats. This allowed the net to suspend straight up and down in the water. Each net would be suspended either from shore or between two boats.

By around 1864, gillnetting had expanded to European, Japanese, and other international fisheries. The boats used by these fisherman were typically around 25 feet long and powered by oars. Many of these boats also had small sails and were called "row-sail" boats. At the beginning of the 1900s, steam powered ships would haul these smaller boats to their fishing grounds and retrieve them at the end of each day. However, at this time gas powered boats were beginning to make their appearance, and by the 1930s, the row-sail boat had virtually disappeared. In 1931, the first powered drum was created by Laurie Jarelainen. The drum is a circular device that is set to the side of the boat and draws in the nets. The powered drum allowed the nets to be drawn in much faster and along with the faster gas powered boats, fisherman were able to fish in areas they had previously been unable to go into, thereby revolutionizing the fishing industry.

During World War II, navigation and communication devices, as well as many other forms of maritime equipment (ex. depth-sounding and radar) were improved and made more compact. These devices became much more accessible to the average fisherman, thus making their range and mobility increasingly larger. It also served to make the industry much more competitive, as the fisherman were forced to invest more into there boats and equipment in order to stay up to date with the current technology.

This type of net was heavily used by many Japanese, South Korean, and Taiwanese fishing fleets on the high seas in the 1980's. Because of their indiscriminate nature, catching near everything in their path, and lack of regulation as well as a growing concern for the plight of whales and dolphins, gillnets were banned by the United Nations in 1993 in international waters.

AQUATIC SYSTEMS OF THE BASIN

The aquatic systems of the Basin are diverse and productive. Already the resources are under pressure from human intervention and this is likely to increase in future. The current population of the Basin is around 500 million, which by 2031 could increase to over a billion, almost half of which could be below the poverty line. Perhaps the greatest single impact at present is for the diversion and storage of water from the river for irrigation. The annual run-off into the Ganges Basin is approximately 469 billion m^3. Of this, an estimated 85 billion m^3 is diverted by barrages, either into canal systems for irrigation and storage or for hydroelectric schemes. Of this diverted water, 60 percent is accounted for by canal projects. Every major tributary has at least one barrage across it. On each northern tributary from Nepal there are barrages at the border region with India. Near Hardwar, the water of the main stem of the Ganges is diverted by a major barrage into the Upper Ganga Canal that was built in 1854 and is largely used to irrigate 3.7 million ha of land with some electricity is also being generated. The main stem, below the barrage is reduced to a very low flow through much of the year.

This system of barrages greatly compartmentalises the ecosystem and certainly presents major barriers to migrations of fish, which may have a general effect on fish distribution as described elsewhere. They represent an artificial demarcation between the upland and lowland systems of the Ganges and, to some extent; they must act as sediment traps. In other cases, the presence of barrages can accentuate problems of pollution by reducing downstream flows for effluent disposal, particularly in the dry season, which can produce chemical barriers to fish distribution. There will be continued pressure to increase the area of croplands under irrigation within the lowland areas, as it is a principal means whereby crop production can keep up with population increases in future. A significant proportion of water diverted for irrigation is returned to the river although at a lower quality. The influence of micro-pollutants, which are related to the extent of the use of agro-chemicals, is uncertain but is likely to increase as the pressure to intensify agricultural production proceeds.

In the Indian sector alone, more than 150 000 km^2 of the Ganges Basin is irrigated using some 85 000 m^3 of river water and 49 500 m^3 of groundwater but this, as in most irrigation systems, has led to extensive problems of soil salinisation. As a result the salt load of the returning irrigation water over 6.3 million tonnes of salt are estimated to be added to the water annually. However, observations on conductivity levels in the main river suggest this has yet to have major effects on the salt concentration of the river as a whole.

The largest barrage of all is Farakka. This was completed in 1975 and was built without consultation with the downstream user state, Bangladesh (or East Pakistan as it was until 1972), the border of which is 17 km downstream. The barrage is designed to regulate river water discharge and to divert a major part of the dry season flow along the Hooghly Canal towards the Bhagirathi-Hooghly and Calcutta rivers. There are also other barriers beyond the canal on the Baghirati itself. In 1995 a formal water sharing agreement was made between India and Bangladesh that agreed to minimum downstream flows and ensures annual meetings to discuss issues between the states. The impact of the Farakka Barrage complex upon the system would seem to include a progressive reduction in the significance of the hilsa fishery upstream of the Barrage as the route from the estuary via the Bhagirathi arm has been impeded. A further possible effect is that the reduction in flow down the Ganges (Padma) below Farakka allows greater incursions of tidal saline water into the southwestern region of Bangladesh, thereby reducing suitable land and water for agriculture.

It is, however, difficult to substantiate in this dynamic fluctuating tidal environment with great annual variations, mainly because of a lack of historical, baseline data. This indicates the importance of good pre and post implementation studies around such structures. As it is, the estuarine hilsa fishery of the Bangladesh delta seems to be maintained.

Within the delta of Bangladesh the principal process of compartmentalisation is not of the river itself but of the floodplains to facilitate the cultivation of rice. Some 40 percent of the floodplain has been modified by empolderment for flood control and irrigation. This has led to a compartmentalisation of distribution of fish particularly the migratory "white fishes" which includes all the major carps. As elsewhere in the Ganges Basin the catches of major carp have declined markedly in Bangladesh, for example, the major carp portion of the catch on Seimanganj floodplain declined from 66.4 percent in 1967 to 13 percent by 1984.

The compartmentalisation of the floodplains may well have contributed to this. A systematic investigation of the impacts of flood control and irrigation schemes on fisheries showed that under conditions of complete flood control reductions could be 81 percent. Under controlled flooding or partial empolderment there is no significant difference in catch with averages of 100-110 kg ha^{-1} inside and out. What does happen, however, is that there is a reduction in bio-diversity within polders of 19-25 percent but, most significantly, a reduction in migratory species up to 95 percent with the main emphasis being on the small floodplain resident species or black fishes. Given the fishing effort that can be deployed on these species in compartments during the dry season, their regular resurgence indicates very resilient life cycle strategies. Up to 59 percent of annual variance in catches can be due to effort under these circumstances.

To contrast this loss of species and valuable elements of the catch as well as to boost recruitment into local populations, major interventions are being undertaken on the floodplains of Bangladesh to redress the situation, either by direct stock enhancement with fry or juveniles or through habitat improvement and restoration. In the largest scale intervention, 60 000 ha were stocked with 50 million carp fingerlings, a mixture of indigenous major carps, Chinese and common carps, over five years which resulted in an incremental fish production of 20 811 tonnes. There is also evidence that recruitment has been augmented in the wider river populations. Stocking interventions, however, may be accompanied by undesirable social consequences including the further exclusion of poor and vulnerable groups from participation in a fishery that has increased in value. As well as enhancement, recruitment may be increased by such processes as habitat restoration. An exercise in habitat restoration which focussed on the clearing of silted channels connecting floodplains to the main river channel increased the proportion of migratory species caught subsequently, including major carp from 2 percent of the catch to 24 percent, and increased the yield 6-fold, from 1860 kg to 11 384 kg per year throughout the study area. In these ways, some of the impacts of compartmentalisation of the floodplains are being addressed in the delta in a continuing fashion but many problems remain, particularly regarding management and governance issues.

In the upland areas of the basin in Nepal, the greatest impacts are said to be due to erosion and increased sediment load from deforestation and from the need to impound water for hydropower generation. The extent of forest removal and increased erosion, however, is difficult to assess. There is evidence that deforestation is a long-term, historical process that may not have accelerated greatly in recent times. It has also been shown that for a basin the size of the Ganges, the sediment delivery ratio is less than 10 percent and that consequently the main channel carries only a modest amount of sediment from the mountains and that, consequently, anthropogenic influences in the mountains have only a limited impact on the plains. It is possible that most of the sediment in the main river comes from storage places and channel erosion.

The other feature of upper basin use is the harnessing of the rivers for hydropower. Nepal has a great potential for hydropower but as yet only 0.27 percent of its assessed potential is being employed.

The rather scattered nature of its own population renders micro projects and run-of-the river schemes good options for domestic generation but the export potential of electricity to northern India and beyond is a commercial imperative. Generating electricity on this scale seems to involve large storage dams which have a number of negative impacts including, as a barrier to fish distribution, possible resettlement of displaced people, disturbance of fragile environments both at the site and downstream and the need to provide roads and services.

The upland rivers certainly have a significance in the rural economies, which has been under-estimated, and it is, therefore, essential that proper evaluations be carried out on the advantages and disadvantages. Under ideal circumstances run-of-the-river projects can avoid many of the environmental disadvantages of storage dams but cases can be seen in Nepal where all the water of the river passes down the adduction tunnel with negligible flow remaining between inlet and outlet. This can provide as much a barrier to fish and navigation as a dam wall. Proper planning and management is required.

Other anthropogenic effects in the basin include pollution. At present, it is of local significance and largely a feature of the Lower Basin where urbanisation and industrialisation are proceeding. The Ganges Basin is reported to carry some 200 tonnes of biological oxygen demand (BOD) per day gross pollution.

However, it is still relatively localised and focussed on urban centres including Hardwar, Kanpur, Varanasi and Diamond Harbour near Calcutta. In addition, the national capitals of New Dehli and Dhaka both have significant impacts. The dry, low water season poses the greatest problems of dilution and dispersal of pollutants. Probably the worst section is between Kanpur and

Allahabad largely due to discharges from the industries of Kanpur, which include tanneries, metalworking and dairies. This appears to be related to the decline in catch of fisherfolk from 30-40 kg to 15 kg per day downstream of the town. Elsewhere, although loads can be quite high, the river disperses them quite rapidly. Even at Varanassi, the main effluent plume is confined to near the city and the river recovers some 20 km downstream. Nevertheless, national governments are concerned and the government of India has been implementing the Ganga Action Plan to start cleaning up the river and preventing it from becoming worse whilst in Bangladesh, the Global Environment Fund is promoting pollution protection. At least at Patna, general water quality has recently improved with, for example, BOD being cut by 75 percent over a decade.

In general, the status of the Ganges Basin reflects the transitional nature of the economies of the constituent states. In developed countries, the greatest problems are commonly gross domestic and industrial pollution and abstraction whilst in developing countries it is frequently degradation of the basin through inappropriate land use, erosion and habitat loss. All of these processes are occurring in the Ganges and probably provides a number of case studies and lessons to learn for basins currently occupied by poorer states as their economic condition improves with development.

There is, however, no overall management concept for the international basin as a whole. There is no international basin management authority such as exists on the Mekong with the Mekong River Commission and no regular fora for making basin-wide management decisions or for regularly sharing information. Critical information on hydrology and satellite imagery can often be classified for security reasons. Using basin-wide data can add dimensions to predictability and management options. For example, using upstream Indian rainfall data in relation to time series of Bangladesh catch data can help predict the likely impact on fish yields three years later. Taking into account all data on the fishes, the fishing communities and development needs it is possible to plan for fisheries management across Asian river basins including the Ganges. A basin-wide approach to management is essential for the economic and environmental sustainability of such large river systems.

ECOSYSTEM-BASED MANAGEMENT

In the recent years, the fishery science started internalizing the notion that management by single species and only by controlling fishing is illogical and contrary to ecological realities. A new approach entered the domain of fisheries management under the name of "Ecosystem-based management". So what the new approach should be about? Let's consider the main elements of an ecosystem affected by commercial fishery (fishery ecosystem) and their changes in space, time and character. The first element is the *time factor*. Any aquatic ecosystem is a dynamic, pulsating, and ever-changing macro-organism. Thus, trends and fluctuations that have been occurring in the ecosystem throughout history must be taken into account. The second element embraces all the changes and physical, biological and chemical forces, including upstream and coastal pollution, imposed upon the ecosystem by *climatic variations* and the various *human activities*. Whatever happens with inshore and bottom habitats, and the water, affects the ecosystem's biota (*i.e.*, all living things). The third element is made up of the *relationships* between the various species occupying the ecosystem at all stages of their life. This runs from bacteria and phyto-plankton up to top predators, with special attention to the managed "target species", physical factors affecting them, their food, prey and predators. Last, but not least is *fishing*, which apart from massive removal of marine organisms from the system, also may influence the size, age and genetic composition of the fished populations. The fishing itself is influenced by the market and the socio-economic context of fishing people and their communities, and of fish consumers, as well as by the industries and technologies involved. No doubt, *controlling fish harvests is not enough to ensure sustainable fishery and healthy ecosystems*.

"The vastness of linkages between species and critical habitats in a coastal area requires comprehensive management of all its parts". An appropriate approach should incorporate institutional arrangements and cultural factors to provide for better analysis and prediction. The list of the factors other than fishing that affect the size, composition and well-being of populations of marine organisms would be extensive. Here're some examples:

Further expansion of dead anoxic zones, 150 of which were reported in 2003 in bays and semi-enclosed seas, some of which extended up to 27,000 square miles, may well be a greater peril than overfishing. A slowly killing mycobacteriosis epidemic was affecting the condition and size of the striped bass population in Maryland and Virginia and in the heavily polluted Chesapeake Bay. According to some estimates, the word's seabirds consume 70 million mt of food as compared to the 80-90 million mt of global marine landings, and the amount of fish eaten by marine mammals worldwide is several times the worldwide ocean fish harvest.

Climatic fluctuations and separate events cause boom-and-bust sequences among fish populations and other marine organisms. There may be umpteen sorts of association between the physical and biological elements. For example, studies show that the main factors critical for salmon abundance are distance to and size of oceanic areas with temperature and food critical to the survival of juveniles, such as the abundance of krill, which is their main food. Pollution of rivers and estuaries and longshore development that destroys inshore habitats, affect the reproduction of many fish species whose spawning and nursery areas are in inshore waters.

FISHERIES SUSTAINABILITY

In most coastal areas, rehabilitation of marine habitats and restoration or protection of essential upstream environments and coastal hydrology functions are essential to future fisheries sustainability. The fundamental message is that for such sustainability we must restructure general coastal management. Environmentalist NGOs and everyone involved in trying to secure sustainability of habitats must realize that there has to be a balance between the needs of the people and the conservation of living resources, and both sides need to be prepared to compromise. The environmentalists' and some scientists' advice that by only controlling fishing they'd achieve sustainability is counter-productive when it diverts attention from other, often more significant factors.

Only an intelligent *analysis and synthesis* of the interaction of all important elements involved can enable understanding of the *problematique* of ecosystem management and searching for solutions. We have to accept that the dynamics of the system is so complex that even the best existing models are of limited use for forecasting the outcomes of actual management actions. In this general context, "*ecosystem-based fishery management*" must address the particular elements. Those that require intervention must be defined and managed respectively, while the others, including those that are beyond the management's powers, must be taken into account. Flexibility and tailoring management for each specific fishery and area should become the rule. There's no rational way to impose a policy that doesn't take all the above into account, or a single "common" policy for different fisheries, ecosystems and fishing cultures. We have to analyze separately each fishery ecosystem to see, for example, if TACs are the right methodology to use or other options such as effort control should be considered.

Unfortunately, some authors and "green" zealots are trying to hijack the concept of "ecosystem management" by representing it as care-taking of the fishery that it doesn't affect the ecosystem. While paying lip service to other factors, they focus on commercial fishing in a way which stigmatizes it as the only factor affecting ecosystem that needs tackling and as the main villain to be constrained or eliminated in order to "protect" marine ecosystems. Their interest doesn't lie in cleaning up habitats and keep fisheries prosperous, but elsewhere. Where? – This depends on who's the "father" (or sponsor) of the particular "stewardship". The institutional and human inertia is strong and many NGOs as well as the fishery management system worldwide abound in scientists and environmentalists who preach sustainability, but keep talking of "saving the oceans" just by shutting down fisheries or by closing off major sea areas to fisheries, and hardly even mention the whole set of other vectors, whatever may be their relative roles in each separate ecosystem. They still ignore physical and biological-ecological environmental factors and upstream and marine pollution from industrial, municipal, and agricultural sources. They close the eye on inshore habitat destruction, and disregard or play down non-human predation, nowadays enhanced by the extensive protection bestowed upon marine mammals.

In fact, however, ecosystem management approach places fishing in its proper context, so it can't be blamed anymore for everything that happens in the sea. Ecosystem management doesn't allow managers and politicians to keep turning a blind eye to pollution, environmental variations, habitat destruction, etc. It carries fishery management from the current over-simplification into more complex concepts. This obviously worries some scientists and managers, who are concerned that with "ecosystem management" they'd be sailing in unknown waters and lose their control over fishing rights and quotas.

Doubtless, it won't be easy, but the point is that to manage fishery ecosystems using the existing methodologies, as they've practically been since the mid 20th century, is like looking under a streetlamp for a lost key, because that's where the light is, instead of looking for it where it may be-however dark is the place. The new approach may also require to get used to the fact that for a long time to come we won't be able to put numerical values to all acting factors, a current custom that sometimes borders on numerology, and relearn to talk also in qualitative terms and perhaps to employ fuzzy logic.

While ecosystem-based management is coming of age, institutional inertia and obstinacy, human conservatism, fear of the unknown and need to admit own misconstruction of the realities of the coastal, marine and living resources, impedes its application. The notion of positive development stems from the recognition that, climatic conditions permitting, sustainability of both, marine capture fisheries and marine aquaculture, can only be achieved by integrated and balanced approach to ecosystem management involving upstream and downstream influences, fisheries, and any other human activities that affect coastal and offshore habitats. There is a growing understanding of the complexity of the systems to be managed and of the futility of simplistic approaches, such as those of the still prevailing fishery management based on inadequate science and its mathematical models.

FISH HEALTH INSPECTION AND CERTIFICATION

Commercial fish farms that ship fish alive, especially interstate or internationally, are often required to conduct inspections for fish diseases. Each state, federal, and international jurisdiction may write and enforce its own unique fish health regulations.

Before shipping any live product anywhere, it is imperative that you make sure that you understand the regulations of all the legal jurisdictions involved, that you have complied with all of the requirements and obtained all of the needed permits, and that your shipment (number, species, dates, etc.) is exactly according to those permits and regulations. Any interstate violation of fish health regulations can put you and your farm at risk for severe penalties (including felony convictions) and fines under state laws and the Federal Lacey Act.

Importation of fish into the United States is under the jurisdiction of the U.S. Fish and Wildlife Service (USFWS) (salmon, trout, and other salmonid fish species), the U.S. Department of Agriculture-Animal and Plant Health Inspection Service (USDA–APHIS), and the destination state. Federal and state permits may be required prior to shipment into the United States. Fish are often restricted to entry through certain ports where federal inspectors are available. Before importing fish, consult with the USDA-APHIS-National Center for Import and Export (NCIE) for advice and with the agriculture and wildlife agencies of the destination state.

The NCIE can also assist you with exports to other countries. For other shipments of live animals, sometimes even within your state, be sure that you have complied with all state aquatic animal health regulations. The regulatory agency with jurisdiction over aquatic animals varies from state to state, and you may need to consult both the state agriculture agency and the state wildlife agency.

Extension fish health experts, other farmers, regional aquaculture centers, state aquaculture coordinators, and fish health testing laboratories may bc valuable sources of information and assistance; however, penalties for errors can be severe, so careful shippers always confirm their regulatory knowledge with an official representative of the regulatory agency.

CURRENT STATUS OF AQUACULTURE GENETICS

DOMESTICATION AND STRAIN EVALUATION

When wild fish are moved to aquaculture settings, a new set of selective pressures comes into play that often changes gene frequencies. Thus an organism better suited for the aquaculture environment begins to evolve. This process, termed domestication, occurs even without directed selection by the fish culturist. Domestication effects can be observed in some fish within as few as one to two generations after removal from the natural environment. In channel catfish (Ictalurus punctatus) an increased growth rate of three to six per cent per generation was observed.

The oldest domesticated strain of channel catfish (89 years), the Kansas strain, has the fastest growth rate of all strains of channel catfish. Although most domesticated strains usually perform better in the aquaculture environment than wild strains, there are exceptions, *e.g.*, wild Nile tilapia, Oreochromis niloticus, and rohu, Labeo rohita, grow better in the aquaculture environment.

The explanation for this appears related to a lack of maintenance of genetic quality and genetic degradation in domesticated strains. Poor performance of some domestic tilapia is related to poor founding (parental) lines, random genetic drift, inbreeding and introgression with slower growing species, such as O. mossambicus and slower growing strains such as Nile tilapia from Ghana. Channel catfish strains differ in growth, disease resistance, body conformation, dressing percentage, vulnerability to angling and seining, age of maturity, time of spawning, fecundity and egg size. Strains of rainbow trout, Oncorhynchus mykiss, show similar variability. Domestication of farmed shrimp (penaeids) has been relatively slow compared to that of finfish.

This can be attributed to use of wild broodstock and postlarvae, a lack of understanding of shrimp reproductive biology for domestication of the species and perceptions of low potential for genetic improvement. Current reliance on wild broodstock is risky and negates the opportunity to enhance disease resistance (as well as other production traits) through selective breeding. Efforts to domesticate broodstock are now hampered by endemic disease challenges; however, recent collaborative research between the Commonwealth Scientific and Industrial Research Organisation (CSIRO) in Australia and the shrimp culture industry has resulted in successful captive breeding of Penaeus japonicus. Economic analysis has demonstrated that domesticated broodstock are more cost-effective than wild broodstock and that reproductive performance of domesticated P. monodon can match that of wild broodstock of a similar sise.

Use of established, high-performance domestic strains is the first step in applying genetic principles to improved aquaculture mana-gement. Strain variation is also important, since there is a strain effect on other genetic enhancement approaches, such as intraspecific crossbreeding, interspecific hybridisation, sex control and genetic engineering.

INBREEDING AND MAINTENANCE OF GENETIC QUALITY

It is as important to prevent production losses due to inbreeding as it is to increase production from genetic enhancement. This applies especially to species with high fecundity, *e.g.,* Indian and Chinese carps, where few broodstock are necessary to meet demands for fry and broodstock replacement. The detrimental effects of inbreeding are well documented and can result in decreases of 30 per cent or greater in growth production, survival and reproduction.

INTRASPECIFIC CROSSBREEDING

Intraspecific crossbreeding (crossing of different strains) may increase growth rate but heterosis (differences between offspring and parents) may not be obtained in every case. Increases of 55 per cent and 22 per cent in growth rate of channel catfish and rainbow trout crossbreeds, respectively, were achieved using this technique. Chum salmon crossbreeds, however, have shown no increase in growth rates compared with parent strains.

Common carp crossbreeds generally express low levels of heterosis, however, those that exhibited positive heterosis are now the basis for carp aquaculture in Israel, Vietnam, China and Hungary.

The crossing of common carp lines in Szarvas, Hungary is a good example of the relative success of crossbreeding. During the last 35 years, more than 140 crosses have been tested.

Three were chosen, based on ~20 per cent improvement in growth rate (and other qualitative features), compared to parent and control carp lines, for culture purposes. Now approximately 80 per cent of common carp production comes from these "Szarvas" crossbreeds. Production of gynogenetic female lines and gynogenetic sex-reversed inbred male lines from common carp with the best combining ability was an important part of the Hungarian crossbreeding programme. A higher heterosis was expected from crossing inbred lines, but the growth rate of F1 crossbreeds was only 10 per cent higher than controls.

Kirpichnikov successfully produced a new strain of cold resistant carp - the "Ropsha" carp - for cold zones in northern Russia using local carps and Siberian wild carps from the River Amur. The Czech Republic also had a national common carp breeding programme and improved growth rates with crossing, "South Bohemian" x "Northern mirror" carp and the "Hungarian 15" x "Northern mirror". Likewise in Israel, over 20 years of crossing common carp strains revealed that the cross using the strain "DOR-70" and the Croatian line "Nawice" produced fast growth, and this is now one of the most popular crosses for Israeli carp production.

In Indonesia, strain development using artificial gynogenesis and sex-reversal resulted in 10 common carp inbred lines, which were later used for crossbreeding. In Vietnam, eight local varieties of common carp, along with "Hungary", "Ukraine", "Indonesia" and "Czech" strains are maintained, with significant heterosis observed in F1 generations of crossbreeds.

Double crosses among Vietnamese, Hungarian and Indonesian strains have subsequently been used for carp selection and crossbreeding programmes throughout Vietnam.

The Vietnamese x Hungarian common carp crossbreed is particularly popular, due to fast growth and high survival rates under different production conditions. Under various rice-field conditions, growth rates of different strains of Nile tilapia and their crosses showed that the crosses were superior to pure Senegal strains.

Breeding programmes are also under development in several countries for the Java or silver barb, Barbodes (formerly Puntius) gonionotus, another economically important fish species in Southeast Asia. The Bangladesh programme used three strains: "Bangladesh", "Thailand" and "Indonesian".

The growth rate of females from six crosses was 23 per cent higher than the average growth rate of the parent strains. Even higher growth rates (35 per cent improvement) were found in the three crosses using the Thailand strain as either the sire or dam. In the Vietnamese breeding programme, six different strains were used to produce a population ideally suited for culture.

Cross-breeds of different strains of European catfish, Silurus glanis, are characterised by outstanding adaptability under warmwater holding conditions and mixed diet feeding regimes.

In addition, crosses with the walking catfish, Clarias macrocephalus, have shown improved resistance to Aeromonas hydrophila infections. Studies on "domestic a" × "domestic b" channel catfish also showed greater heterotic growth rates than "domestic" × "wild" crosses (Dunham and Smitherman, 1983. The same was found in rainbow trout crosses. Strain mating incompatibilities, however, can occur and can impede fry output.

PARASITIC DISEASES

ARGULUS (FISH LOUSE)

Symptoms: The fish scrapes itself against objects, clamped fins, visible parasites about 1/4 inch in diameter are visible on the body of the fish.

The fish louse is a flattened mite-like crustacean about 5 mm long that attaches itself to the body of fish. They irritate the host fish which may have clamped fins, become restless, and may show inflamed areas where the lice have been. With larger fish and light infestations, the lice can be picked off with a pair of forceps. Other cases can best be done with a 10 to 30 minute bath in 10 mg per litre of potassium permanganate. Or treat the whole tank with 2 mg per litre, but this method is messy and dyes the water.

ANCHOR WORM (LERNAEA)

Symptoms: The fish scrapes itself against objects, whitish-green threads hang out of the fish's skin with an inflamed area at the point of attachment.

Anchor worms are actually crustaceans. The young are free swimming and borrow into the skin, go into the muscles and develop for several months before showing.

They release eggs and die. The holes left behind are ugly and may become infected. The anchor worm is too deeply imbedded to safely remove. Treatment can best be done with a 10 to 30 minute bath in 10 mg per litre of potassium permanganate. Or treat the whole tank with 2 mg per litre, but this method is messy and dyes the water.

BLACK SPOT - BLACK ICK (DIPLOPSTOMIASIS)

Symptoms: The fish, very irritated, scrapes itself against objects, appears as small black specks or smudges on the body and around the mouth, and if heavily infected may experience blood loss. Black Spot or Black Ick is rare in aquariums. It is generally seen in outdoor ponds, especially those with mud bottoms, but it can be introduced when adding new fish into the aquarium. Fish that are most readily susceptible are the Silver Dollar, Piranha, or other fish of these types. In general it does relatively little damage to the fish, even if they are heavily infested. The disease is caused by a parasite (larval trematode) that burrows into the skin of a fish where it forms a cyst that is about one millimeter in diameter. It has a complex lifecyle that in order to survive, requires fish eating birds or animals, snails, or fish at different stages infected with the disease. Black Spot is generally easy to cure. Either treat with salt baths or there are a number of commercially available treatments and preventatives.

ERGASILUS

Symptoms: The fish scrapes itself against objects, whitish-green threads hang out of the fish's gills. This parasite is like the anchor worm, but is smaller and attacks the gills instead of the skin. Treatment can best be done with a 10 to 30 minute bath in 10 mg per litre of potassium permanganate. Or treat the whole tank with 2 mg per litre, but this method is messy and dyes the water.

FLUKES

Symptoms: The fish scrapes itself against objects, rapid gill movement, mucus covering the gills or body, the gills or fins may be eaten away, the skin may become reddened.

There are many species of flukes, which are flatworms about 1 mm long, and several symptoms that are visible. They infest gills and skin much like ich, but the difference can be seen with a hand lens. You should be able to see movement and possibly eye spots, which is not found in ich. Gill flukes will eventually destroy the gills thus killing the fish. Symptoms of a heavy infestations are pale fish with drooping fins, rapid respiration, glancing off aquarium decor, and/or hollow bellies. Treatment can best be done with a 10 to 30 minute bath in 10 mg per litre of potassium permanganate. Or treat the whole tank with 2 mg per litre, but this method is messy and dyes the water.

NEMATODA

Symptoms: Worms hanging from the anus.

Nematodes (threadworms) infect just about anywhere in the body but only shows itself when they hang out of the anus. A heavy infestation causes hollow bellies.

Lighter infestations usually cause no problems with the fish. Short of destroying the fish, which is easier, two treatments have been suggested. First treatment; soak the food in parachlorometaxylenol and give the fish a bath or treat the aquarium with 10 ml per litre. The bath should last for several days. Second treatment; find special food containing thiabendazole as a nematode (threadworm) cure and hope the fish will eat it.

LEECHES

Symptoms: Leeches are visible on the fish's skin.

Leeches are external parasites and affix themselves on the body, fins, or gills of the fish. Usually they appear as heart shaped worms (they are just curled up) attached to the fish. They are usually introduced to the aquarium via plants or snails. Since leeches are sucking and borrowing into the surface of the fish, removal with forceps can cause great damage, if not death, to the fish. If the fish is bathed in a 2.5 per cent solution of salt for 15 minutes, most of the leeches should just fall off. Those that do not will be affected enough to remove with forceps with minimal damage. Another treatment is to add Trichlorofon at 0.25 mg/l to the aquarium. Live plants should be removed and treated with potassium permanganate at 5 mg/l before replanting.

URONEMA MARINUM

Symptoms: Skin scraping, pale discoloration, loss of colour, weight loss, dehydration, flashing, and rapid breathing.

The saltwater parasite, *Uronema marinum*, is a free-living ciliated protozoa that can cause fatal infections in marine fish. It is an opportunistic feeder that normally eats on bacteria, but when the immunization of a fish is low it will attack, invading the fish's muscles and internal organs. This infestation is often the result of the introduction of a new fish, overcrowding, and poor water quality resulting from a high organic load in the aquarium.

This parasite is difficult to identify as the symptoms can also be indicative of other parasitic and bacterial problems. However, it can be debilitating and ultimately fatal to a variety of marine fish including Tangs, especially the Yellow Tang, Angelfish species especially those in the genus *Centropyge*s, Seahorses, many species of Butterflyfish, yellow headed Jawfish, and others. The best way to avoid the problem is to keep your current tank free from infestation. Quarantine all new fish for a period of three weeks, improve the water quality of the tank, and reduce the stress level in the aquarium by reducing the number of fish and incorporating places for fish to hide and rest.

There are several types of medications that can be used to treat infected fish:

- Medications such as Malachite green, Copper Sulfate, or Methylene blue. Use caution and be sure to follow the manufacturers instructions.
- Freshwater bath—place infected fish in the freshwater bath for a period of three minutes or until the fish shows signs of stress.
- Low salinity (hypo salinity) treatment—lower the salinity in the quarantine tank to a specific gravity of 1.011 and maintain at this salinity for 21 days. Do not use this treatment with invertebrates or especially sensitive fish such as sharks and rays.
- *Nitrofurazone*: An antibiotic that has some antiparasitic action, and can be helpful when used along with formalin dips.

6

Products of Aquaculture

AQUACULTURE PRODUCTS

Aquaculture product expansion has placed increased requirements on quality and food safety by consumers and regulators. There are few questions about the evident nutritional benefits of consuming fish but, by the end of the 1980s, developed countries arrived at the conclusion that classic fish (and food) inspection procedures, based on the analysis of samples of the final product and on generic hygiene measures, were not enough to provide the necessary level of protection to consumers. A preventive system called HACCP (hazard analysis critical control point) was adopted, and governments started to shift their regulations to HACCP-based systems. If one word has to be chosen to explain HACCP, it is "prevention."

The growth of aquaculture has led to significant changes in how its products are perceived and marketed. In becoming an important contributor to the markets for seafood, aquaculture is increasingly subject to safety mechanisms and controls, such as the statutory hazard analysis critical control point (HACCP) methodology in certain developed regions. As both safety and trade regulations are harmonized at international levels, quantitative risk assessment and traceability will become integral components of aquaculture management. Developing countries have increased their share of the seafood export market to nearly 50 per cent of global trade, a significant portion being represented by aquaculture products (shrimp, salmon, molluscs, etc.), a percentage that should increase with the continued expansion of the sector.

The long-term viability of aquaculture development will be market driven, accounting for consumer demand and the capacity to adapt to the structure and legislative demands of the target markets. Important externalities will affect the production sector in achieving its goals, including the topics of sustainability, traceability and interactions with the environment. The development of schemes of best practice that incorporate the addressing of such issues, alongside quality schemes and safety management, will aid the production sector to achieve its goals. Discriminatory tariffs on trade should be avoided, particularly given the increasing importance of aquaculture as an export-earner and contributor to food security. The sectoral response to such developments must include a better understanding of the complex issues faced by the production and marketing of consumer food products within highly competitive markets, where the assumption of responsibility is essential, both for the product and the actions taken to produce it.

At a global level, aquaculture is one of the fastest growing food production sectors (9.6 per cent/yr in the last decade), a fact that will ultimately change the way that fish is perceived as food. A key element of this observation is the change in the supply opportunities for fish and fish products from a wild source to a cultured one. Aquaculture product expansion has placed increased requirements on quality and food safety by consumers and regulators. There are few questions about the evident nutritional benefits of consuming fish but, by the end of the 1980s, developed countries arrived at the conclusion that classic fish (and food) inspection procedures, based on the analysis of samples of the final product and on generic hygiene measures, were not enough to provide the necessary level of protection to consumers. A preventive system called HACCP (hazard analysis critical control point) was adopted, and governments started to shift their regulations to HACCP-based systems. If one word has to be chosen to explain HACCP, it is "prevention."

An important point within the global market is the growing importance of international agreements that involve food (and fish) safety aspects. About 40 per cent of all fish produced are traded internationally, which means that there is a search for common criteria that facilitate or permit clear rules for compliance. The tendency is, therefore, to move towards the harmonization of national regulations, meaning that such regulations could assure an equivalent level of food protection to consumers. This is a relatively easy concept to understand, but very difficult to implement and validate in practice. In turn, this increases the importance of internationally accepted guidelines, recommendations and standards, such as those of the Codex Alimentarius. The provisions related to food trade of the General Agreement on Tariffs and Trade (GATT) Agreement compound this tendency, and all of these aspects are interlinked. The extent of regional and international trade in aquaculture products is difficult to analyse because trade in many aquaculture products is not yet well documented in all of the main producing countries. Aquaculture contributes primarily to domestic consumption but, at an international level, important trade has developed for a number of aquaculture products, and this has been one of the principal driving forces for aquaculture development in many developing and developed countries.

INTERNATIONAL AGREEMENTS ON FOOD SAFETY

While there is no common agreement on food or fish safety at the international level, agreements do exist on food trade that have implications for food safety and quality matters. The important one regarding food safety is the Agreement on the Application of Sanitary and Phytosanitary and Measures (SPS), which introduces two very important concepts. The first is that "to harmonise sanitary and phytosanitary measures on as wide a basis as possible, Members (countries) shall base their sanitary or phytosanitary measures on international standards, guidelines and recommendations, where they exist... ".

The same agreement establishes that, regarding food safety, "international standards, guidelines and recommendations" to be taken as reference will be those "established by the Codex Alimentarius Commission (CAC) relating to food additives, veterinary drug and pesticide residues, contaminants, methods of analysis and sampling and guidelines of hygiene practice".

The second important point of the SPS Agreement is the adoption of the criterion that risk assessment shall be the basis for determining the appropriate level of both sanitary and phytosanitary protection. In this case, the SPS Agreement imposes a strong condition on national regulations that have not yet accommodated this criterion. Current HACCP-based regulations include a "hazard analysis" step, but risk assessment, which should be a part, is performed only qualitatively, and regulations do not refer to risk assessment explicitly. It is clear that the future development and evolution of regulations will be in this direction. All developed countries and a large number of developing countries have taken up regulatory HACCP-systems. These concern the safety of fish and fish products, which include products from aquaculture, and it is assessed that approximately 65 per cent of the total international fish trade is performed under such regulations. The greatest exception concerns the Japanese market, which accounts for about 24 per cent of the total fish market (demand), but which has no HACCP regulations as yet.

FISH SAFETY IN THE EUROPEAN COMMUNITY

The Member States of the European Community (EC) are subject to both European and national legislation, and considerable developments in laws affecting aquaculture have occurred in the last decade. These are separated into two clear sections that concern the final product and the aquaculture process.

The HACCP-based regulation for fish and fish products in the EC introduced the concept of "own health checks" (HACCP in terms of the EC regulation relative to fish products), and it is the Council Directive 91/493/EEC that laid down the health conditions for the production and the placing on the market of fishery products. EC aquaculture production follows Council Directive 91/67/EEC, which concerns conditions of stock health and health management and which, in practice, generated a "family" of regulations. For instance, Council Directives 93/54/EEC, 95/22/EC and 95/70/EC have successively amended

Directive 91/67/EEC. This "family" of regulations is aimed basically at controlling fish diseases; they are only, in a second instance, complementary with the HACCP-based regulations aimed at preventing human diseases. The regulations related to aquaculture production do not imply accomplishment of the fish safety regulations, however. Although there are no contradictions between either regulation "family", their complementarity and coordination will surely increase with time.

FISH SAFETY IN THE UNITED STATES OF AMERICA

The basic regulation in the United States of America, which makes mandatory the implementation of the HACCP system in fish and fish products, is that of the United States Food and Drug Administration (FDA,5).

This regulation does not mention fish obtained from aquaculture specifically, but this source is included implicitly in the definition of "fish" ("where such animal life is intended for consumption") and in other paragraphs of the regulation (*e.g.*, when "drug residues" is listed as one of the possible hazards to be accounted for). However, the regulation is specifically centred on processing, and does not apply to aquaculture production in itself.

As a general conclusion, one can say that such regulations are structured around a core principle that defines the HACCP-based system, and that this basic principle is complemented by a number of regulations that can be defined as "regulatory standards". Each species and product may be defined within a "family" of regulatory texts that outline the minimum requirements for each particular case or procedure.

Fish and fish products from aquaculture are included, either explicitly or implicitly, in such HACCP-based regulations. However, whereas the HACCP system is well defined for processing steps or procedures once the fish has been harvested, the application of the HACCP system to the entire aquaculture production chain is not so clearly established within the regulatory environment. Developments in this area can be anticipated.

Further evolution of the fish safety regulations may be expected with the introduction of new technical, scientific and operative knowledge and, in particular, regarding mandatory quantitative risk assessment.

All matters that relate to safety, quality and trade rely on the ability to identify and trace the product, but the traceability of food remains as an enormous problem to be solved. In its broadest sense, a traceability system would provide and allow access to all information concerning a product. In the case of aquaculture, this could mean following the history of a production batch from the final point of sale back to the hatchery. This would mean assuring knowledge of the steps encountered in distribution, packaging, processing and harvest, including the growing conditions, the feeds and therapeutic agents used, the broodstock parentage and any other relevant information.

Clearly, such a huge amount of information cannot be encoded physically with the product, but needs to be accessible via coding and linking to appropriate databases. The ability to trace and isolate a problem with a product is essential to any safety system, but the concept of traceability is much more than this, since it affords assurance, generates trust and, very importantly, can allow the use of important information for marketing and sales. Information is a new measure for adding value and will increase in importance.

AQUACULTURE EXTENSION

Aquaculture contributes significantly to the rural economy of most of the Asian and other developing countries by providing part- and full-time occupation to the farmers, fishermen and landless agricultural labourers. India and other developing countries of the South Asian region are endowed with ample water resources in the shape of freshwater ponds and tanks for fish culture, but these are not under effective and optimum utilization in spite of highly developed available technologies.

Research results have shown excellent production potential as well as economic viability of such technologies, but until these technologies are successfully transferred to the beneficiaries, the desired objective cannot be achieved. Like agriculture, aquaculture is also an agroclimatic-based technology which, when developed in one agroclimatic region, may need modifications and refinement for adoption to another region. Agriculture extension thus involves not only the extension of aquaculture technology, but also certain levels of adoptive research in a particular field environment before it is launched for large-scale extension.

It is a two-way education process in which both scientists and farmers contribute, receive and interact with the involvement of extension workers as a link between the two and a catalyst as well (model). In other words, it is a non-formal adult education programme for educating and training the rural mass to acquire suitable fish farming skills and capabilities with a view to boosting fish production efficiency and the socio-economic condition.

Aquaculture extension is basically an educational process by means of which scientific and technological knowledge of aquaculture is carried to the farmers to upgrade their existing operation and farm management skills. The philosophy behind this process is to change the altitude, enhance the skill and knowledge of the fish farmers to upgrade their aquaculture practice. It also aims at binging maximum possible unutilized and under-utilized water areas under modern fish culture operation so as to raise the standard of living of the fish farmers through improving productivity and profitability. Apart from achieving its own target its overall objective is also to signifiantly contribute towards rural development by improving rural economy, creating additional gainful employment opportunities, fighting malnutrition and preventing rural exodus.

LAUNCHING AQUACULTURE TENSION PROGRAMME

Any aquaculture extension programme is designed based upon broad national consideration to achieve national goals and targets viz-a-viz local considerations to achieve short-term objectives such as application of composite fish culture in undrainable ponds to improve the aquaculture production level. A local aquaculture extension programme is relatively more definite in terms of scope and target. Like any other aquaculture extension programme, there should be three sequential steps for the dissemination of fish culture technology in undrainable ponds. They are as follows:

- Programme planning
- Programme implementation
- Programme evaluation

Programme Planning

While planning the dissemination of fish culture technology one should always bear in mind that the programme should be a self-regenerating production endeavour and once it is stimulated should continue on its own with a changed attitude and active participation of the recipient. This involves situation-specific strategies. The main components of programme planning are pre-adoption survey of the area, situation analysis, setting programme goals and finally designing strategies in a sequential manner.

Village Survey

Fish culture is basically a rural farming system and hence village survey is the most common method for identification of the difficulties faced by the farmers and to find out the scope and suitability of a specific technology needing to be transferred. The main objective is to get an overall picture of the village and the villagers, their attitude, values, together with their socio-economic conditions and also to locate and assess the available freshwater resources.

It also helps to identify the local institutions, village leaders, progressive farmers, school teachers, village level workers in order to design the most feasible extension strategy and also to establish a permanent rapport to strengthen the extension services. At micro level it provides information about the socio-economic conditions of individual fish farmers, the pattern of fish farming and the technological gap. The village selected for the survey should be such that it may represent the locality. Regular contact with important and progressive farmers of the village should be maintained. They should be informed about the objectives of the survey proposed to be undertaken.

Interviews with these persons will provide an overall picture of available natural and human resources and possible areas for development. Finally detailed relevant information may be collected from individual pond owners/ fish farmers and fishermen through personal interviews/questionnaires.

Resource Inventory

- *Availability of water resources:* Various types of water resources are available for fish culture but usually all of them are not fully utilized. Large, medium and smaller types of water bodies are generally available in villages which may be suitable for fish culture. Many small water bodies are found fully shaded by large marginal trees and thereby lying unproductive. Some unconventional types of water areas with potentiality for intensive aquaculture are also available.
  Canal/road and village side small and large ditches, pits emerging due to construction of mud houses etc., are some of the unnoticed and untrapped potential aquaculture resources suitable for seed production and short-term fish rearing.
  Low-lying and swampy areas which are formed naturally due to human activities are also potential sites for undrainable ponds for fish culture.
- *Availability of human resources:* It is a well-known fact that the majority of the people in developing countries live in villages and most of the rural population depend upon agriculture, aquaculture, livestock farming and other allied activities for their livelihood. Human resources are the vital inputs in rural aquaculture development. Rural areas have vast potential of unutilized or underutilized human resources for both men and women, which can be effectively utilized in operating aquaculture.
- *Identification of possible constraints:* A village survey also offers an excellent opportunity to identify various constraints in the background of which an appropriate strategy can be suitably designed.

Financial

Farmers usually do not have surplus funds big enough to be diverted towards reclamation and renovation of existing watersheds as well as construction of new ponds. Initial expenditure for fish culture over fish toxicant, fish seed and supplementary feed is itself a considerably big amount to be exclusively borne by farmers themselves without any credit support. As such, possible sources for mobilizing credit facilities may be identified.

Improper Water Area Distribution Pattern

Like land distribution pattern, major water areas are usually found in the possession of medium and big farmers who bother least about fish culture and concentrate themselves mostly on agriculture, while small and marginal farmers have minimum water holdings at their disposal with adequate manpower potential to be utilized. In some areas most of the water bodies are vested to village institutions, local administrative bodies, etc.

Lack of Technical Knowhow

Several seasonal and perennial ponds without any proper embankments are found lying fallow in a derelict condition due to ignorance and lack of technical knowhow. In some cases farmers fail to follow-up the prescribed package of practices strictly and land themselves in a state of financial turmoil and lose confidence in the viability of newly developed fish farming technologies.

Lack of Stocking Materials and Other Material Inputs

Fish farmers usually face the biggest problem of unavailability of quality fish seed for stocking their pond. Paucity of quality fish seed in the locality force the farmers to stock their ponds without any consideration to proper stocking size, density, species, ratio, etc.

At times, they procure riverine fish seed which is usually mixed with the seeds of predatory and weed fishes. Other material such as fish toxicants are usually localised in its availability. All such problems are also vital for deciding area specific extension strategies.

Marketing Problem

It is a general practice that the fish is sold to middlemen at the pond site who invariably pay lower prices. Due to the perishable nature of the commodity and fear of exploitation by the fish wholesellers, farmers prefer to sell the crop at their pond/farm sites even at relatively lower rates. Information related to marketing practices will add to the scope of the extension programme so that farmers may be educated in marketing management to avoid such exploitation.

Lack of Transport and Efficient Communication System

In remote villages of India and many developing countries where fish culture technology needs to be extended, proper transport and communication facilities are lacking.

Social and Administrative Problems

Ponds remaining unutilized and lying in derelict conditions are common sights in rural areas in spite of a certain level of fish culture know how available with the farmers. In most cases such conditions exist due to family rivalry and non-cooperation among the members of the owners especially when the water areas are under multi-ownership. Poaching and deliberate poisoning of the ponds to destroy the crop are also serious social problems.

In some areas fish culture is supposed to be of a low-caste profession, thus many efficient upper-caste prople remain reluctant to come forward for this venture. Local administration such as Panchayats and Block Level Development Departments are also not always suitable geared enough to ensure rural aquaculture development.

Setting Programme Goals and Planning

In the light of resource inventory and possible limitations suitable target groups may be identified, programme goals may be set up and accordingly suitable extension strategy may be planned. Without such an early insight and planning, the programme may not have firm and realistic footing. Although the fish farmers are the usual target of any fish culture extension programme, all the fish farmers may not be suitable to be involved for immediate participation.

Target groups may be selected on a number of criteria including farming practice, production level, income, education, cultural background, nature, reputation in the society, initiative, liable to change their attitude, etc. Selection of suitable communication channels is also very important. Data collected during the pre-adoption survey provide the necessary information for such selection. After these selections, programme goals may be set up. Goals indicate the direction towards which the programme is oriented. It also provides reference level for evaluating the programme achievements. Examples of goals in such extension programme may be on the following lines:

- Improving the socio-economic uplift of fish farmers and raising the standard of living;
- Bringing 100 per cent of the available undrainable ponds for composite fish culture.

Programme Implementation

"Plan the work and work the plan" is an appropriate term for any extension programme. Once the plan is laid, all possible efforts should be diverted to ensure that responsibilities are carried out, schedules are followed arid activities accomplished as per the plan. Although the strategies and planning of the aquaculture extension programme are situation specific, some general steps may be cited as follows;

- Through heavy flow of information using mass media, publications, individual and mass contacts, etc., awareness and interest should be created among fish farmers.
- One or two demonstration centres may be set up and the technology of composite fish culture and seed production in undrainable ponds may be demonstrated to maximum possible farmers to let them realize the case of operation, production potential and profitability.
- A set of suitable farmers should be selected initially and be motivated and guided enough so that they strictly follow the different package of practices as per the schedule.
- Proper steps may also be taken to make available the critical material inputs at the pond/farm sites and if the programme permits, subsidy should be given as a token of initial attraction.

- If the availability of quality fish seed is found to be a limiting factor, fish seed may be distributed free of cost or at concessional rate to the farmers at the initial stage. Proper attention may also be paid for extending fish breeding and seed rearing programmes.
- Facilities for proper monitoring of water quality and fish health may be extended through the participation of nearby laboratories.
- Periodical netting for growth check/health inspection should be strictly followed and supervised.
- Self-explanatory/pictorial instruction booklets dealing with basic steps of composite fish culture, control breeding of common carp, techniques of pituitary gland collection, induced breeding of major carps, hatchery operation for carps, nursery and rearing pond managements, techniques of fish seed transport, etc., may be prepared, explained and distributed among farmers.
- Ad hoc training courses should be organized at the demonstration sites on different aspects of fish culture and fish seed production for participating and other interested farmers. Exhibition programme/Fish Farmer's Day should be organized time to time at different places in which live specimens of all the six carp species, other culturable air-breathing fishes, harmful predatory species, weeds, fish feeds, fish toxicants, etc., should be shown and the objective and goal of the programme may be exhibited through models, charts, posters, etc.
- At times a team of a few farmers may be selected on the basis of their leadership quality and performance and sent to visit important aquaculture centres, farms, research institutions, etc.
- Individual contacts through home visits is a very effective extension method. The extension worker must be very clear in this objective during the visit and must do sufficient preparation with regard to subject matter information he is going to deliver to the fish farmer and family members.
- Evening is the most suitable time for organizing an assembly of farmers. Necessary details about practices to be followed the next day may be explained to them during such assemblies. Teaching aids may be used to make the communication effective.

Programme Evaluation

Programme evaluation is the process to determine the extent of success of the executed extension programme in the light of present objectives. It is an important management function in order to ensure effective implementation of the programme. It also helps in the identification of the deficiencies and weakness of the programme so that proper corrective measures may be taken to make it more useful in its future course. Programme evaluation can be conducted once a year or at a specific period of the programme and finally at the concluding phase of the programme. The process of evaluation also depends upon the nature of the programme. A short term and less extensive localized extension programme may be evaluated by the extension workers themselves through the analysis of progress reports, field records, questionnaires, etc. However, broad-based and elaborate extension programmes can be evaluated by specialists in association with the extension workers to determine the effectiveness and impact of the extension programme.

It is convenient to fractionate the whole programme into smaller components for effective and easy evaluation. Fractionation may be done as follows:

- Resources (financial, personnel an material) made available.
- Objective of the programme in clear terms.
- Phases of the programme (evaluation should also be done phase-wise).
- Data collection from records and tabulation.
- Selection of ways, means and methods for the collection of data/information from participating, non-participating fish farmers, village youths, prominent persons of the locality, etc.
- Sample selection
- Collection of data/information from target and non-target groups
- Tabulation of data
- Data analysis and interpretation of results

To measure the degree of success, certain values have to be associated with the information. Increased fish production level, profit through increased fish yields, knowledge of modern management techniques, fish breeding, fish seed rearing, increased number of ponds/water areas in the area, etc. are some of such measurable values for programme evaluation.

IMPORTANT CONSIDERATIONS

- Extension services can be made most effective by making the people understand, accept and adopt the new technology, as it is very much clear from statistical data that people remember 10 per cent of what they hear, 40–50 per cent of what they see and hear and 90 per cent of what they see, hear and do.
- Maximum potential for development of undrainable pond fish culture lies in developing countries where the prevailing literacy level is lowest.
- The personality profile of extension workers is of prime importance for the extension of any rural-based aquaculture technology. He must mould himself enough so that he may become technical by profession, social by temperament and preserve human values and missionary zeal of serving the rural poor. He should be simple, easily approachable and adaptable enough so that he can live among fish farmers. At the same time he must not have any inferiority complex while meeting with specialists and higher bureaucrats.

- The target groups of aquaculture extension programmes are usually socio-economically backward rural masses having a low level of literacy and technical knowhow and are reluctant to bring about quick changes in their attitude.
- Many extension departments and voluntary organizations are also of the opinion that the gains of socio-economic and technological development progrmames do net reach the rural poor and that the roots of this failure lies in the lack of organization of the poor themselves. Field workers experience very often instances of diversion of financial grants meant for production programmes for the small and marginal farmers into consumption subsidies. Similarly, there are numerous instances where free educational facilities are granted for children of rural poor, but they hardly avail the opportunity as they are treated as helping hands for supplementing family income. Most of the resources allocated for various welfare activities of the rural poor, however, are diverted into activities totally unrelated to mass welfare. The fact is that the poor are not only poor but are disorganized and hence they have very little influence in the process of decision-making and implementation of the programme. Under such conditions the concept of community fish farming may also be considered as an effective and ideal method for organizing at least a section of the rural poor/fish farmers/fishermen in cooperative and productive communication.
- Support services and credit facilities are the two important factors which play major roles in the aquaculture development programme. Lack of appropriate support services and proper credit facilities are the major drawbacks.
- Effective institutional support to provide the necessary technical services needed by the extension programme, such as site selection, pond designing, fish health check, pond environment monitoring, etc., are vital for programme implementation.

CULTURE-BASED FISHERIES

The text also includes provisions for culture-based fisheries which are taken to mean capture fisheries which are maintained by stocking with material raised within aquaculture installations. However, this definition is too narrow to cover the range of management practices collectively known as enhancements, and for the purposes of this document the following *working definition* for culture-based fisheries is given as: - Activities aimed at supplementing or sustaining the recruitment of one or more aquatic species and raising the total production or the production of selected elements of a fishery beyond a level which is sustainable through natural processes. In this sense culture-based fisheries include enhancement measures which may take the form of: introduction of new species; stocking natural and artificial water bodies; fertilization; environmental engineering including habitat improvements and modification of water bodies; altering species composition including elimination of undesirable species, or constituting an artificial fauna of selected species; genetic modification of introduced species.

FOOD SECURITY

Food security exists when all people, at all times, have physical and economic access to sufficient, safe and nutritious food to meet their dietary needs and food preferences for an active and healthy life.

Sustainable Development

Sustainable development is the management and conservation of the natural resource base and the orientation of technological and institutional change in such a manner as to ensure the attainment and continued satisfaction of human needs for present and future generations. Such sustainable development (in the agriculture, forestry and fisheries sectors) conserves land, water, plant and animal genetic resources, is environmentally non-degrading, technically appropriate, economically viable and socially acceptable. Responsible development of aquaculture, including culture-based fisheries, in areas under national jurisdiction "States should establish, maintain and develop an appropriate legal and administrative framework which facilitates the development of responsible aquaculture."

General responsibilities. In order to promote, support, and regulate an efficient and responsible aquaculture sector, States should establish, maintain and further develop an appropriate administrative and legal framework to ensure that responsible aquaculture practices are introduced and implemented within their national jurisdiction.

Designated authority. States should designate or establish an authority or authorities competent, empowered and capable to effectively promote, support and regulate aquaculture and culture-based fisheries. Appropriate institutional linkages with other authorities such as those concerned with agriculture, rural development, water resources, environment, health, education and training and many others, should also be established. These linkages may have to be expressed in legislative form.

Legal framework. States and their aquaculture authorities should ensure that the aquaculture sector is adequately regulated and protected by legal instruments such as laws, regulations, orders, agreements, etc. which set forth the responsibilities, rights and privileges of aquaculturists in a manner which is consistent with the current and potential aquaculture practices and with those applied to comparable activities.

ENFORCEMENT OF AQUACULTURE LEGISLATION

Understanding and enforcement of aquaculture legislation. States and their aquaculture authorities should ensure that all applicable legal instruments including laws, regulations, orders, etc. are conceived in such form as to be readily understood by those undertaking activities within the aquaculture sector, are adequately communicated to them, and finally that these legal instruments are enforceable and enforced.

"States should promote responsible development and management of aquaculture, including an advance evaluation of the effects of aquaculture development on genetic diversity and ecosystem integrity, based on the best available scientific information."

General responsibilities. States should, through their competent authorities, and in partnership with all interested actors of civil society, promote development of environmentally sound and sustainable aquaculture well integrated into rural, agricultural and coastal developments, raise awareness of the general public of the benefits of aquaculture practices for enhanced food supply and income generation, and support efforts aiming at responsible actions of aquafarmers and all those concerned or associated with aquaculture.

Advance evaluations of genetic effects. Genetic effects may arise from the interaction of farmed species with wild species and can be caused by the use of introduced species and by species that have been domesticated, or genetically modified by an aquaculture breeding programme or other technologies. Undesirable genetic effects may include:

- Contamination of native gene pool through interspecific hybridization,
- Degradation of native species through flux of exotic genes that are 'less fit', either by means of hybridization or, hypothetically, by gene transfer,
- Loss of native species, or change in species composition through competition, predation, or habitat degradation.

  Advance evaluation of genetic effects should include a risk assessment that examines:

- The probability of an escape from the culture system,
- Survival of the organism if it does escape,
- The organism's reproductive capability in the wild, both with itself and with other species,
- The ability of specific genes from the species to be transferred to native species.

An assessment of the potential damage should be made and, in essence, the key question is "does the organism present a danger to the ecosystem or to important species in the area ?" In relation to genetic technologies and risk, the change the technology imparts to the organism should be evaluated rather than the technology itself. If certain technologies such as gene transfer are associated with high levels of uncertainty as to their effect on the organism or the environment, then more testing, stricter regulation and monitoring should be followed.

For advance evaluation (*i.e.*, pre-impact assessment) to have significant and practical meaning, there should be predetermined standards, *i.e.*, acceptable limits of impacts. Present knowledge of many ecosystems and their genetic diversity is often very incomplete, especially in many developing countries and tropical regions. Setting standards for allowable genetic "effluent" is still very difficult, because of the scarcity of information on the effects of aquaculture/ wild animal interaction, survival of aquaculture escapees, and their impact on ecosystems. These acceptable levels of impact will not only be necessary to make advance evaluation meaningful, but they may also serve as guidelines or benchmarks for the monitoring of the ensuing aquaculture development. The Precautionary Approach to Fisheries Management requires the establishment of such benchmarks, as well as contingency plans when acceptable levels are breached. Collaboration among environmental scientists, aquaculture experts and development planners will be required in many cases to successfully apply precautionary approaches and to implement advance evaluations. However, it should be noted that even the models available for the assessment and prediction of ecological impacts of aquaculture wastes are usually highly site-specific, often quite sophisticated, and, in most cases only applicable in temperate regions. Therefore, cost-effective and rapid assessment methods are needed which can be applied easily in tropical environments and developing countries. In general terms, environmental scientists can help by clearly distinguishing between actual and hypothetical environmental hazards resulting from aquaculture practice.

"States should produce and regularly update aquaculture development strategies and plans, as required, to ensure that aquaculture development is ecologically sustainable and to allow the rational use of resources shared by aquaculture and other activities."

Aquaculture Development and Support Planning

Aquaculture development and support planning. In many countries there is a continued need for aquaculture and planning authorities to produce and regularly update comprehensive plans for promoting, supporting, regulating, and reporting on the aquaculture sector.

The plans should encompass all relevant aspects of support and management of the industry. Given the possible contributions of aquaculture to enhanced food supply and rural development in many countries, it may be very useful to design aquaculture development plans with due consideration of existing plans and efforts aiming at food security, sustainable agriculture and rural development.

Development planning will involve, possibly in consultation with all interest groups, the setting of policies and objectives, determining and implementing of actions required, monitoring the sector's performance, and adjusting the aquaculture development plan. Good collaboration among those concerned will help identifying the type of data and information necessary for monitoring and planning.

Ideally, an aquaculture development plan should reflect the responsibilities of all parties concerned, possibly in accordance with the principles and guidelines of the Code of Conduct for Responsible Fisheries. Clearly, an aquaculture development plan should also suit the needs of individual States while taking into account the social and economic conditions in which aquaculture practices are or will be carried out.

Where appropriate, the mechanisms for consultation of aquafarmers and other private sector representatives as well as the involvement of local authorities and communities could be further specified in aquaculture development plans, in particular with regard to their role in the implementation and enforcement of the Code's principles and guidelines. Assuring appropriate and responsible use of land and water resources. It should be ensured by aquaculture and planning authorities as well as by aquaculturists and investors that aquaculture activities are sited in locations which: are suitable for sustainable production and income generation; are economically and socially appropriate; prevent or minimize conflicts with other users of resources, and do not create undue externalities; respect nature reserves, protected areas, and critical or especially sensitive habitats.

Government authorities should also ensure that the privileges and needs of the aquaculture sector are recognized and respected by other users of land and water and, in particular, that aquafarms are not exposed to external environmental threats resulting from activities in other sectors that reduce quality and quantity of water, nutrient and biological resources required. Where applicable, zoning or site regulations should be specified to conform with the requirements of plans for regional development, river basin or coastal area management, and their respective authorities.

As generally done with agricultural and forestry activities, aquaculture and fishery enhancement practices should also be duly considered in planning and management of inland and coastal resources.

In many countries there is a need for collaboration and capacity building in many areas relevant to resource use planning for aquaculture development, including application of resource assessment and monitoring methodologies, cross-sectoral resource use planning and management, environmental accounting, and mechanisms for conflict resolution and involvement of stakeholders in decision-making processes. Those who have special responsibilities and skills with respect to the development of aquaculture and other sectors should work together for the sustainable use of resources, maximizing wherever possible their complementarities.

For example, when reviewing and implementing water resource use policies, it may well be worth considering benefits of existing and future aquaculture and inland fisheries developments. In particular, water allocation and pricing policies should address potential economic and social benefits of incorporating aquaculture and fishery enhancement practices. At the same time, most efforts aiming at the conservation of water resources and/or the protection of aquatic environments can and should be beneficial to fisheries and aquaculture, thereby increasing food security and generating some net economic gain for a local community, or in certain cases, a positive contribution to the economy of a given country.

CULTURE-BASED FISHERIES AQUACULTURE

In contrast to aquaculture which involves the cultivation of aquatic life within controlled environments or the commercial production of certain aquatic species by managing the major part of their life history under strict control, culture-based fisheries increase production in natural environments by controlling a part of the life history of certain species and transplanting or releasing their seed or fry into the open waters. The juvenile fish, which are produced in hatcheries and are released into fresh-, brackish-, or marine waters, are allowed to propagate or grow on natural foods until they reach harvestable size.

Unlike aquaculture, culture-based fisheries are not limited by land or population pressures and do not have to modify or manage the culture system to approximate the natural environment since it uses the natural environment itself. Also, unlike aquaculture, harvests are uncertain and returns are more difficult to predict as the "release of fish to sea involves the number harvestable ultimately being reduced to 1-15 per cent of the number released, depending on species and locality".

Nonetheless, culture-based fisheries have been increasingly resorted to as means of enhancing the fishery resources, replenishing natural stocks whose populations have declined through over-exploitation or environmental degradation, or simply maximizing the productivity of a water body, be it an open bay, a coastal lagoon, or a freshwater reservoir. Their adoption or application on an expanded scale has in fact been identified as a high-priority item in the agenda for development in a number of regions in the world in the coming years. The major types of culture-based fisheries include: (i) sea/ocean ranching as practised in Japan with various marine finfishes and the Kuruma prawn, and in the Pacific Ocean and Baltic Sea with salmon; (ii) coastal lagoon farming, as exemplified by valliculture in the Mediterranean; (iii) stocking and restocking in freshwater lakes and reservoirs; and (iv) floodplain fisheries management.

Sea/Ocean Ranching

Sea ranching, in which juvenile fish are released to grow unprotected on natural foods in marine waters where they are harvested at marketable size, takes advantage of the sea as a large aquatic pasture. While it is practised in slightly differing forms around the world, in its distinctive form, young fish are released to the sea through ponds, enclosures or channels, and allowed to grow until they are ready for the market.

Salmon

Sea ranching of salmon is widely practised in countries such as the United States (mainly Alaska, Washington and Oregon), Canada, Japan, USSR, New Zealand, Iceland, and the Baltic countries.

The most widely used species for ocean ranching is the Pacific salmon (Oncorhynchus sp.) because of the success of its artificial spawning coupled with the rapid growth and the high degree of its homing precision that makes it amenable to harvesting at maturity. The practice of salmon ranching was initially carried out more than 100 years ago in the Nordic countries where salmon smolts were released into the Baltic Sea area.

Hatchery production of salmon in the United States dates back to 1872 when the US Fish Commission established its first hatchery on the McCloud River in California. This was followed by the first private salmon hatchery on the Rogue River, Oregon in 1877. The technology for artificial propagation of salmon was introduced to Japan in 1877 and to western Canada in 1884.

In Japan, while salmon hatching and release have been carried out for nearly 100 years, the results were not considered very successful. It was only with the new method of intermediate feeding and timely release of salmon fry, starting in 1961, that significant advances in salmon production have been achieved. Most Japanese hatcheries use gravel-lined raceways to incubate larval salmon (alevins) and practise short-term feeding of newly-emerged chum fry on artificial diets.

Pink and chum salmon are raised in equal numbers in Soviet hatcheries using technology similar to that of Japan. In the Pacific Northwest, chinook and coho salmon are the predominant species used in hatcheries where raceways and ponds for feeding juvenile fish are prominently featured.

In Alaska, floating raceways in protected saltwater bays had to be innovated to suit Alaskan conditions. Canada's salmon farming industry has grown tremendously in the last five years as a result of its successful hatchery operations. Canadian hatchery workers have pioneered development and application of spawning and egg incubation channels which provide favourable environmental conditions for eggs and alevins. These techniques have been applied most successfully to sockeye, pink, and chum salmon. The spawning channels produce fry with a capacity for rapid growth and high survival because they simulate high-quality natural spawning beds. Incubation channels, shallow raceways with a silt-free gravel bed, enable the alevins from artificially fertilized and hatched eggs to complete their development in a natural substrate.

Salmon fry that need to be raised in low-salinity or fresh water before release into marine waters (*e.g.*, sockeye, coho, and chinook) have to be transferred from the incubation system to feed-lot systems consisting of tanks, raceways, ponds, floating pens, or a combination or modification of these. The newly emerged fry are given frequent feedings of high-protein diets containing vitamin supplements and certain fats.

Other Species

Aside from the widely popular Pacific salmon, other marine species have been used for ocean ranching in Japan. The Japanese National Culture-Based Fishery Project, which was started in 1953, has succeeded in developing techniques for mass seed production and subsequent release of juveniles of commercially valuable fish/shellfish, such as the Kuruma prawn, red sea bream, blue crabs, sole, flounder, and yellowtail, into the Seto Inland Sea. The Project has now reached the stage where the fishermen themselves carry out the release of fry and manage the propagation work. The Project, based on its earlier successes, has been expanded on a national scale with the construction of a large number of Prefectural Centers for Culture-Based Fisheries all over Japan. The Ezo abalone, Haliotis discus Hanai, is one of the most important species for ocean ranching in Japan, with about 1.15 million hatchery-produced juveniles released in 1980 in rocky areas of the Sea of Japan around Hokkaido. Likewise, the giant Ezo scallop, Patinopecten yessoensis, is propagated in the Okhotsk Sea, the Nemuro Channel, and in other areas of the Hokkaido coast.

COASTAL LAGOON FARMING

Extensive lagoon farming is a traditional activity in various European and Mediterranean countries with a production of about 10 000 t/y. Among the Mediterranean countries, Italy has the largest areas of brackish water that are exploited for lagoon farming of commercially valuable species like Anguilla anguilla, Mugil cephalus, Liza aurata, Liza saliens, Liza ramada, Chelon labrosus, Dicentrarchus labrax, and Sparus aurata. Restocking of the valli/coastal lagoons is carried out by anadromous, natural, and annual migration and by artificial restocking. Catches from the valli primarily depend on the quantity of fingerlings which entered the lagoons some years before and on the quantity of fingerlings stocked.

Valliculture

Valliculture, as the practice is called in view of its being fish culture-based, is one of the most ancient forms of aquaculture in the Mediterranean region. Its origins date back to the first rudimental fish pond and fattening systems used along the Adriatic and Tyrrhenian coasts. This technique was developed by the upper Adriatic populations to exploit the seasonal migrations of some fish species from the sea into the lagoon and delta areas which were more suitable for their growth. The fish returned to the sea because of altered environmental conditions (temperature) of the sea or for reproduction. To exploit these periodic movements, large brackish areas were enclosed to prevent the fish returning to the sea and complex permanent capture systems, fish barriers, were developed consisting of barriers in the channels communicating with the sea to catch the adults. Later, from the simple ponding of fry freely entering the lagoon from the sea, came a man-made seeding of fry fished elsewhere and introduced into the basins to be reared for a few years.

The successful management of coastal lagoons using hatchery-bred fry of important species is thought to hold considerable promise world-wide for expansion in the future.

Brush-park Fisheries

Brush-park fisheries are a traditional form of low-technology aquaculture which is practised in coastal lagoons and brackish waters in many areas of the world. The brush parks are constructed in a variety of forms and sizes, but basically a brush park consists of an inner core, or concentric circles, of densely packed tree branches or other material surrounded by an outer, more substantial wooden framework which is fished periodically, usually by encirclement.

In coastal lagoons the most sophisticated forms of brush-park fisheries occur in Benin. There are several basic types of brush-park fisheries in Benin where they are known collectively as "acadjas". They vary from each other in configuration, construction and fishing characteristics. The acadjas are constructed of tree branches with harder woods forming the peripheral structure and with soft wood branches with many ramifications forming the interior of the acadja. Acadjas are preferentially placed in shallow, quiet waters of no more than 1.5 m depth. Fishing is accomplished by surrounding the smaller types of acadjas with a net after which all of the branches are removed and the net is then pursed. For the larger brush parks the surrounding net is fished in a step-wise process by gradually moving the net inwards as branches are removed until the fish are concentrated in small areas and can thus be removed with traps, baskets, and hand-nets. After fishing, the used branches are replaced and new ones added as necessary.

Elsewhere in Africa, brush-park fisheries are found in coastal lagoons in Nigeria, Côte d'Ivoire, Ghana, Togo, and Madagascar. Elsewhere in the world, brush-park fisheries are found in Negombo lagoon in Sri Lanka and have been introduced in some Mexican lagoons.

Stocking of Inland Waters

Among the most advocated and widely used practices for the management of the fisheries resources of lakes and reservoirs throughout the world is the stocking and recapture technique. Species stocked are either indigenous or exotic and either herbivorous, carnivorous, or omnivorous. In the developing countries of the tropics, where the production of more fish for food is the goal of any fisheries activity, high-yielding herbivores, detritivores and plankton feeders (like tilapia and carp), are commonly stocked in lakes and reservoirs. This is in direct contrast with the developed countries in temperate zones where carnivores like trout and bass are stocked in lakes and reservoirs to support the recreational or sport fishery.

The major lakes and reservoirs in ASEAN countries, for example, have been stocked with hatchery-produced fingerlings of a wide variety of species including tilapia, various Chinese and Indian carps, milkfish, mullet and catfish.

In China, around 65 per cent, or 1.4 million ha, of reservoir area out of a total of two million ha are farmed and managed like fish ponds. They are stocked with fingerlings and following a period of growth, mature fish are harvested. The species used are the common carp, grass carp, silver carp, bighead carp, mud carp, and Wuchang bream (Megalobrama amblycephala) in varying densities and species ratios. Reservoir fisheries in China are almost entirely dependent on the regular release of large quantities of fingerlings because the Chinese carp do not spawn outside their area of origin. In many instances, therefore, stocked fish make up over 90 per cent of fish catches from reservoirs.

In Israel, many water storage reservoirs (10-40 ha area, 5-7 m depth) originally constructed for irrigation, are now stocked with a range of species including tilapia, mullet and carp, usually grown in intensively managed polyculture systems. Sri Lanka's large perennial reservoirs and small seasonal tanks constitute the country's main freshwater resource. These are regularly stocked with fingerlings of the different carp species (Indian, Chinese, and common) which are produced at Government-run hatcheries spread all over the country. Since the Indian and Chinese carps do not spawn in the reservoirs/tanks, they have to be stocked regularly to sustain the fisheries. There is thus large pressure on the State to accelerate its seed production programme to keep up with the large demand for fingerlings. It is interesting to note that unlike the Philippines and Thailand, for example, where fish seed production is largely a private sector undertaking, the Sri Lankan Ministry of Fisheries is primarily responsible for the production of carp fingerlings for distribution among the various reservoirs and seasonal tanks.

In Cuba, most of the freshwater fish is produced from seeding of the country's small to medium size reservoirs; in Colombia, some dams are stocked and restocked with mullets.

African small reservoirs are stocked mainly with tilapias and Clarias. In Zambia the most important species for stocking reservoirs are Tilapia melanopleura, T. macrochir, T. andersonii, and Haplochromis spp. In Zimbabwe the fish most commonly used for dam stocking are Oreochromis mossambicus, O. macrochir, T. rendalli, and several exotics like black bass and rainbow trout.

STATUS, TRENDS AND ISSUES IN FISHERIES AND AQUACULTURE

We suspect that living in true harmony with the natural world, in a manner sustainable over the long run, is something no modern human society has yet learned how to do. The survival of the natural world, however, and likely our survival as a species, depends on our learning to do this. It will be a unique experience in human history.

MARINE FISHERIES RESOURCES AND POTENTIAL

Reported global production of marine capture fisheries increased from 17 million tons in 1950 to about 80 million tons in the mid-1980s, oscillating since then between 78 and 88 million tons (excluding discards), representing 60 per cent of the overall fisheries production including aquaculture in 2001. The annual rate of increase of marine catches decreased to almost zero in the 1990s, indicating that the world oceans have reached their maximal production under the present fishing regime. An estimated 25 per cent of the major marine fish stocks for which information is available are underexploited or moderately exploited. Stocks or species groups in this category represent the main source for the potential expansion of total marine catches. About 47 per cent of the main stocks or species groups are fully exploited and are therefore producing catches that have reached, or are very close to, their maximum sustainable limits. Thus, nearly half of world marine stocks offer no reasonable expectations for further expansion.

Another 18 per cent of stocks or species groups are reported as overexploited. Prospects for expansion or increased production from these stocks are negligible; and there is an increasing likelihood that stocks will decline further and catches will decrease, unless remedial management action is taken to reduce overfishing. The remaining 10 per cent of stocks have become significantly depleted, or are recovering from depletion and are far less productive than they used to be (or than they could be) if management can return them to the higher abundance levels commensurate with their pre-depletion catch levels. Recovery usually implies drastic and long-lasting reductions in fishing pressure and/or the adoption of other management measures to remove conditions that contributed to the stock's overexploitation and depletion. In most areas, overfishing is certainly a significant factor responsible for the declines. The information available tends to confirm the estimates made by FAO in the early seventies that the global potential for marine fisheries is about 100 million tons of which only 80 million tons were probably achievable for practical reasons. It also confirms that despite local differences, overall, this limit has been reached.

Developed countries are faced with fully or overexploited stocks so their management objectives concentrate on stock rebuilding and capacity reduction, although most countries also have significant aims regarding markets and social conflict. The most urgent objective is to scale fleet sizes so that they become commensurate with sustainable exploitation of the resources. Management plans also increasingly recognise the need for a policy that integrates fisheries with management of the coastal zone or inland waters.

Developing countries tend to concentrate on fisheries development in terms of new resources and technology. Although it is recognised that some stocks are overfished, objectives are concentrated more on enhancing and diversifying fisheries rather than on limiting fishing efforts. This is perhaps because the underlying concern for many countries is the relatively important role fisheries play in employment and food security for some of their poorest people. More specific aims include building infrastructure (particularly for processing to reduce post-harvest losses and increase the value added); fishery enhancement, through restocking; and reducing social conflicts, not only among different fishing groups but also between fisheries and other sectors. The principal policy challenge is to bring the capacity of the global fishing fleet back to a level at which fish stocks can be sustainably harvested. Fisheries based on clearly defined rights of access will need to become more common: experience shows that when these rights are not merely in place but are understood and observed by users, conflicts tend to be minimized. The need is to blend innovation, research, conservation and educational awareness into a goal of aquatic sustainability, and to demonstrate this via successful commercial enterprises.

The "great trends" of the past 20 years in marine fisheries worldwide are the:

- Globalisation of both production and labour in marine fisheries,
- The overcapacity of fleets both nearshore and offshore, resulting in dramatic and widespread declines in catches per unit efforts,
- The movement of the bulk of the world's fishing capacity from the developed to developing countries, resulting in
- Declining catches and economic hardships worldwide in small scale artesanal and inshore fisheries resulting in less fish for poor consumers, and
- Infant success with various property rights schemes and use of protected areas to sustain stocks.

Inland Fisheries Resources and Potential

Four current strategies in the use of inland waters for fisheries can be distinguished:

- Food fisheries based on wild stocks;
- Enhancement of food fisheries in smaller water bodies and reservoirs;
- Recreational fisheries, which are becoming more common in many areas of the world, and, where they develop, tend to supplant commercial food fisheries; and

- Locally very intense exploitation of juvenile or small adults for stocking into other water bodies and/or aquaculture ponds, or for the ornamental fish trade.

In 2001, inland capture from 129 countries reached 8.7 million tons. Regionally, Asia dominates inland capture, accounting for nearly 66 per cent (about 5.8 million tons) of the total in 2001, with Africa second at 24 per cent (nearly 2 million tons). In comparison, the remaining regions are relatively insignificant. The capture from inland waters is very diverse. Apart from fish, other groups included in inland capture that are tabulated by weight are freshwater shrimps and prawns, frogs, terrapins, turtles, crayfishes, mussels, mussel shells, pearl oyster shells, swamp crabs, marine worms, manatees and green seaweeds.

There is an enigma concerning the status of inland fishery resources. On the one hand, it is widely accepted and well documented that freshwater resources and environments are deteriorating rapidly and widely. On the other hand, harvests from freshwater environments are stable or on the upswing among countries that account for 93 per cent of the total inland capture.

The increases are due to three factors:

1. Enrichment of aquatic systems from land based human activities that until now have been a counterweight to habitat losses and pollution;
2. Enhancements of inland fisheries (*e.g.*, stocking, introductions) that increase the output per unit of area; and
3. The increased number of reservoirs with capture fisheries enhancement activities.31 Inland fisheries enhancement methods, often combined with conventional fisheries management practices, are widely applied, and are becoming a central theme in the management of inland waters in developing countries The exact contribution of enhancements to the total inland capture fisheries production is, however, difficult to estimate.

INLAND CAPTURE IN DEVELOPING COUNTRIES

About two-thirds of inland capture is from developing countries and about one-fourth is from China. The countries in Africa and Asia that are among the top 20 in inland capture production also rank low in the Human Development Indices.

This comparison underlines the need to sustain inland fisheries in the countries where inland capture is relatively important in its own right, where inland capture contributes significantly to food security, and where levels of human development are lowest. In contrast, inland recreational fisheries are pre-eminent in most developed countries; however, almost every country with inland waters has some recreational fisheries that often play an important role in subsistence. Indications are that freshwater recreational effort is perhaps one-half of the food fishing effort from a worldwide perspective.

Output from inland food fisheries will continue to increase slowly overall. In the short term most of the gains will be realised where enhancements are already common, namely in Asia. In Africa, where enhancements have been slow to be implemented, output may actually decrease due to overfishing and a deteriorating environment in the most populous countries, but increase as enhancements become more widely implemented. Likewise, in Latin America, inland fisheries enhancements are practiced and are growing, but the demand for inland fish is weak in many countries, while in others degradation of the environment is causing a decline in production.

Degradation of the environment is the main underlying issue, and the consequent loss of fishery habitat is the pre-eminent concern. A closely related concern is that loss of habitat that, along with the intense and widespread exploitation of fishery resources, negatively impacts aquatic biodiversity. In practical terms, the resiliency of resources to fishing, and often the quality of the resources, is lessened. Additionally, in most countries the main challenges to maintaining and enhancing inland fish production and small-scale aquaculture and their associated social and economic benefits are increasing competition for resources and insufficient institutional and political recognition.

Reporting and Statistical Issues in Fisheries

Without reliable statistics, effective fisheries management and policy-making are impossible. During the last decade, financial support for the development and maintenance of national fishery statistical systems have decreased sharply in real terms, while statistical requirements have been increasing dramatically for by-catch and discards, fishing capacity, illegal fishing, vessels authorised to fish in the high seas, economic data (costs, revenues, prices, subsidies), employment, management systems, inventories of stocks and fisheries, aquaculture, etc.

Despite FAO's efforts, the available fisheries data are not fully reliable. The outcome is far from perfect in terms of coverage, timeliness, and quality. Statistics from artisanal and subsistence fisheries are still a concern, and many key statistics are missing, *e.g.*, economic and social data, discards, fishing capacity, etc. The result is that the general trends are probably reliably reflected, but the annual figures and the assessments involve a degree of uncertainty, and small changes from year to year are probably not statistically significant. Illegal, unreported and unregulated fishing is found in all capture fisheries, irrespective of the location, species targeted, fishing gears employed or level and intensity of exploitation. Working with the countries is the only way to improve fishery statistics, primarily to meet national needs with regard to food security and fisheries management. Unlike capture fisheries, the separate monitoring of aquaculture is relatively new in most countries, and often there are less well-established systems of data collection as compared to capture fisheries.

Actual inland capture fisheries production is considerably greater than the amounts reported to FAO. The factor is at least two overall, but may be as high as three in some instances. There is an urgent need for better data on inland fisheries that can be interpreted in both economic and ecological terms. Although the cost of improving inland fishery data collection may be high, failure to fully account for inland capture also is costly in terms of lost opportunities to increase food security and other economic and social benefits from enhanced management of inland fisheries resources.

China's marine capture fishery production for 1995-1999 has been overstated in Chinese statistics submitted to and published by FAO. The chapter states that as a consequence of this, global marine capture fishery production-excluding Peruvian anchoveta-has likely been declining since 1988 rather than remaining fairly constant as indicated by the statistics. According to the authors, this would have led to understating the degradation of world fisheries, and produced unfortunate policy and investment decisions. However, the FAO maintains that despite likely errors in the data sets, the main global trends have not been masked, and that the most important conclusions have emerged, nevertheless. These findings, together with similar ones emerging at regional and national level, have been the foundation for the governance and institutional changes observed since 1990.

Aquaculture Status and Potential

Aquaculture is more akin to farming and animal husbandry than to fishing, as it involves the rearing and management of living aquatic resources in a restricted environment. Tenure of production facilities, and property rights to the produce, are as important to the success of aquaculture as land tenure is to agriculture.

Aquaculture has been developed to serve a variety of purposes:

- Producing high nutritional value food for human consumption;
- Contributing to rural income and employment through farming and related activities;
- Enhancing capture and sport fisheries;
- Cultivating ornamental species for aesthetic purposes;
- Controlling aquatic weeds or pests hazardous to humans or crops; and
- Desalination and other forms of soil recuperation.

Meeting basic human needs for protein foods in the future will be a difficult challenge. Approximately 1.3 billion people live on less than a dollar a day-the cost of a half a pint of beer-and half of the world's population lives on less than 2 dollars a day.

Since 1950, there's been a 100 per cent increase in the per capita demand for fish, a 40 per cent increase for grain, and 33 per cent for wood. If world fish consumption will increase from 16 kg (1997) to 19-20 kg by 2030, total human use of aquatic foods will increase to 150-160 million tons. Capture fisheries can provide no more than 100 million tons, so the bulk of the increase will need to come from aquaculture.

In 2001, 48.4 million tons of aquatic products (including plants) valued at US$ 61.5 billion were produced, with half of the production being finfish. Aquaculture is growing more rapidly than all other animal food producing sectors. Worldwide, the sector has increased at an average compounded rate of 9.2 per cent per year since 1970, compared with only 1.4 per cent for capture fisheries and 2.8 per cent for terrestrial farmed meat production systems. Over half of global aquaculture production originated from marine and brackish coastal waters, while the remainder (45 per cent) was from freshwater.

Aquaculture will soon overtake cattle ranching as a global food resource, possibly signaling a basic shift in diets. Over the last century, the world relied heavily on two natural systems-oceanic fisheries and rangelands-to satisfy a growing demand for animal protein, but that era is ending as both systems are reaching their productive limits. Between 1950 and 1990, beef production, four-fifths of it from rangelands, nearly tripled, climbing from 19 million to 53 million tons before leveling off. Since 1990, there has been little growth in either beef production or the oceanic fish catch. Additional production of beef or seafood now depends on placing more cattle in feedlots or more fish in ponds.

PRODUCTION EFFICIENCIES OF AQUACULTURE

Comparisons of energy and production efficiencies of aquaculture versus an array of fisheries and terrestrial agriculture systems confirm that aquaculture is an efficient mass producer of animal proteins for a crowded, coastal planet. Production efficiencies of edible mass for aquaculture range from 2.5 to 4.5 kg dry feed/kg edible mass compared with 3.0 to 17.4 for conventional terrestrial animal production systems. Beef cattle require over 10 kg of feed to add 1 kg of edible weight, whereas catfish can add a kg of edible weight with less than 3 kg of feed. To produce 1 kcal of catfish protein about 34 kcal of fossil fuel energy is required-lobster and shrimp capture fisheries use more than 5 times this amount of energy. Energy costs for even the most intensive salmon cages are less than lobster and shrimp fishing, but are comparable to beef production in feedlots. Aquaculture also has a comparable advantage in water efficiency, since a comparatively smaller amount of water for aquaculture is required. For example, catfish ponds managed using multiple harvest strategies use 1.50 m3 of water per 1 kg of product while conventional soybean production takes 1.63 m3 of water to produce 1 kg.

In a world of land and water scarcity, the advantages of aquaculture over capture fisheries and other land-based protein production systems to produce low-cost animal protein are clear. However, care must be taken to ensure that aquaculture is not viewed as a panacea-since most analysts agree that aquaculture will never completely replace capture fisheries-and that increased aquaculture production will not be associated with higher rates of environmental damage and harvest rates of forage fish species used to produce fish meals and oils, which would lead to a net loss of fish production (*e.g.*, the capture for fish meals and oils would exceed the amount of fish produced for consumption).

There are many cases where aquaculture expansion has fueled the hope of fragile coastal and inland rural communities that have undergone unprecedented changes in their traditional ways of life. Aquaculture has provided significant multiplier effects on the local economies increasing both direct and indirect personal spending in these coastal communities. And yet, communities in many parts of the world actively oppose aquacultural development because such development is perceived as a threat to local social and/or ecological systems. If intensive aquaculture operations-no matter how advanced technically-have no community roots; and feeds, seeds, supplies, equipment and human expertise are procured from great distances from the sites of production; community opposition will continue to occur. In these cases it is easy to see why some community members view aquaculture development as "all we get is your pollution". Planning for aquaculture development as community development and environmental enhancement must thereby encompass regional planning processes to accommodate aquaculture's vital support industries (inputs), and for the use of aquaculture resources and wastes in agriculture or in environmental enhancement projects (outputs). Regional planning for "ecological aquaculture" developments will have much higher positive impacts on jobs and the environment, and will eventually dissolve the opposition from communities who will see the newcomers as one of their own.

Aquaculture production systems, aquatic environments, and the feeds to produce fish must be environmentally benign and ecologically sound in order for the advantages of aquaculture as a food production system to be embraced and supported. Indeed, the expansion of aquaculture can never be justified on the destruction of the world's capture fisheries. The future protein needs of millions of people-and the sustainability of aquaculture itself-depend on the conservation, good management and recovery of the world's capture fisheries and the environments on which they depend.

FISHERIES AND AQUACULTURE PRODUCTS

Fisheries and aquaculture products are globally important sources of much needed, high quality, aquatic animal proteins, and invaluable providers of employment, cash income, and foreign exchange. Fisheries products are the world's most widely traded foods, with commerce dominated by the developing countries.

Fisheries products are the primary protein sources for some 950 million people worldwide, and are an important part of the diet of many more. In comparison to other sectors of the world food economy, however, the fisheries and aquaculture sectors are poorly planned, inadequately funded, and neglected by all levels of government. This neglect occurs in a paradoxical situation: fishing is the largest extractive use of wildlife in the world; and aquaculture is the most rapidly growing sector of the global agricultural economy.

Our vision is one in which the USAID is a world leader in channeling high quality, "needs directed" kinds of technical assistance in fisheries and aquaculture to developing countries, mainly in the form of capacity-building though education and training opportunities, but also in applied research. Considering the status, priority issues and future trends we have identified in this study, we recommend seven strategic approaches and investments to USAID:

1. USAID needs to substantially increase its programmatic emphasis and enlarge its financial and human resource commitments to global fisheries and aquaculture.
2. USAID needs to play a central role in mobilizing America's considerable human and institutional resources in fisheries and aquaculture to assist developing countries.
3. USAID needs to bridge the "digital divide" to develop solutions to fisheries and aquaculture *issues in developing countries.*
4. USAID should prioritize the improved management of coastal marine and inland fisheries by providing technical assistance to evolve innovative fisheries management schemes in developing countries, including but not limited to, property rights, co-management, and the use of marine protected areas; plus assist in the development of more accurate and reliable fisheries data reporting systems.
5. USAID needs to substantially increase its support to develop more comprehensive, sustainable, ecologically and socially compatible, and economically viable aquaculture systems in developing countries that have the long-term goals of poverty alleviation and food security.
6. USAID should prioritize its assistance to fisheries and aquaculture activities that are more integrated, comprehensive, community-based, and use "systems approaches"-such as ecological and integrated farming/fishing systems research and extension approaches-in both rural and urban settings. The current agriculture emphasis of USAID is on plant commodity research, not on comprehensive, agro/aqua-ecosystems research/extension approaches. We urge the USAID to support long-term, applied research and development that makes expanded use of participatory ecological and social science tools to empower community control of fisheries and aquaculture systems;

and to better integrate aquaculture and fisheries activities into the comprehensive management of natural and social resources of its missions, target nations and regions.

7. USAID needs to develop comprehensive strategic and implementation plans and regular impact assessments of an expanded fisheries and aquaculture portfolio. USAID missions and regions should include fisheries and aquaculture into their strategic plans for the comprehensive management of natural resources-or they will be incomplete-especially in regards to USAID plans for involvement in the issues of water allocation and quality, and plans for the management of marine and inland coastal areas.

We believe our case for increased strategic engagement and substantially increased investments in priority issues of importance to the future sustainability of global capture fisheries and aquaculture by all levels of the USAID bureaucracy (headquarters, regions, missions, etc.) cannot be overlooked. America has an accelerating trade deficit in fisheries products that now exceeds $9 billion per year-a deficit surpassed only by those in oil and automobiles-and the US remains the 4th largest exporter of fisheries products in the world (~$2.8 billion). The lack of US engagement in international fisheries and aquaculture not only compromises America's financial position: and an important part of our Nation's food security is at risk; and our domestic fisheries and aquaculture industries are rapidly losing their competitive position.

TRADED AQUACULTURE PRODUCTS

In 1998, the main internationally traded products from aquaculture were shrimp, salmon molluscs and seaweed. Other species showing strong growth are tilapia, seabass and seabream.

CRUSTACEANS

Marine shrimp is the most prominent product from aquaculture in international trade, and aquaculture has been the major force behind increased shrimp trading during the past seven to eight years. Shrimp is already the most traded seafood product internationally, with about 25 percent or 800 000 mt coming from aquaculture. Since the late 1980s, farmed shrimp has tended to act as a stabilizing factor for the shrimp industry. Therefore, the major crop failures in Asia and Latin America during the past few years (caused by disease problems) have had a significant impact on overall supply, demand, prices and consumption trends.

The major markets are Japan, the United States of America and, to a lesser extent, the EC, while the largest exporters of farmed shrimp are Thailand, Ecuador, Indonesia, India, Mexico, Bangladesh and Vietnam. Demand for shrimp is expected to increase in coming years, where Asian markets, such as China, the Republic of Korea, Thailand and Malaysia, will expand as local economies recover and consumers' demand more seafood. This trend is already reducing the availability of shrimp to traditional importers and will eventually put upward pressure on prices if supplies do not increase.

Trade in crab species has also increased with growing aquaculture production (1997: 165 000 mt). Especially important have been the exports of China (19,000 mt in 1998) to Hong Kong SAR China and Japan.

Finfish

In terms of total aquaculture output, finfish ranks first, with 49 per cent of the total production from aquaculture, of which the major part are carp species, which are consumed locally in the producing countries (mainly China and India). As opposed to shrimp, finfish aquaculture trade appears to be split between species having a high traditional demand and a "quality" image (*e.g.*, salmon, seabass etc.) and convenience products (mainly fillets) of "cheaper" fish species (*e.g.*, tilapia). The following species are the main products that are seen as being important in international trade:

Salmon: International trade in farmed salmon has increased from virtually nothing to more than 600 000 mt (1999) in less than a decade. The traded species are mainly Atlantic salmon and, to a much lesser extent, coho salmon, which accounted for 87 per cent and 12 per cent of 1997 salmon production, respectively. The growth seen in trade has mirrored that of production, reflecting the fact that this activity is concentrated in a few countries that have limited domestic markets (Norway, Chile and the United Kingdom). Norway, whose main market is the EC, is the main exporter of Atlantic salmon. On the other hand, Chile, whose principal markets are Japan and the United States, is the main exporter of coho salmon and the second largest exporter of Atlantic salmon.

As production volumes have increased, so has competition within the market and costs and prices have been driven down. At the current price levels (+/-US$4/kg CIF [Cost, Insurance and Freight) Europe), salmon has become a relatively medium-priced product in international seafood markets.

Trout: Trout is traded internationally at a much lower level than salmon, with exports reaching 135 000 mt in 1998 out of a total production of 463 000 mt.. However, trout production is split, approximately equally, between portion-size production and large trout, and where 70 per cent of global production is in Europe. Consumption, particularly of portion-size fish, is concentrated in trout-producing countries, but Norway, Chile and Finland have been able to farm particular qualities of large-sized, heavily pigmented trout for the Japanese market, primarily as a replacement product for coho salmon. Japanese trout imports reached 60 000 mt in 1998.

Tilapia: Tilapia species have also shown a tremendous growth in output (production of 946 000 mt in 1997), and although international trade is limited, it is growing. This is observed especially between Latin America (Costa Rica, Ecuador and Colombia) and the United States, as well as between Asian producers (Taiwan Province of China, Indonesia and Thailand) and the United States and Japan. There is also modest trade between the EC and producers in Jamaica, Israel and Zimbabwe. The biggest exporter, Taiwan Province of China, supplies Japan with high-quality tilapia fillets for the sashimi market and ships frozen tilapia to the American market. Taiwan Province of China now exports about 35 per cent of its domestic tilapia production and supplies 80 per cent of the United States tilapia imports (1998). Thailand and Indonesia export less than five per cent of their production. Vietnam has also recently entered the world tilapia market, and China exported 500 mt to the United States in 1998.

Tilapia has become the third highest imported aquaculture product, by weight, in the United States (1998 imports of 28 000 mt), after shrimp and salmon. United States imports rose 14 per cent by quantity in 1998. In the long term, tilapia prices are expected to decrease, and this should lead to greater exports to the United States, as well as to Europe, which is seen presently as being undeveloped as a market for tilapia.

Seabass and Seabream: In Europe, the marine fish farm industry of the Mediterranean has focused on the production of European seabass and gilthead seabream, and it intends to copy the success of salmon growers. Production reached almost 90,000 mt in 1999, of which nearly 90 per cent was exported from the country of origin, mainly to Italy and Spain.

The principal exporter was Greece, which exported about 70 per cent of domestic production. Italy was, for a long time, almost the exclusive market for Greek production. However, as a result of market development efforts, some 15 per cent of Greek exports are now going to "new" markets (*e.g.*, the United Kingdom, Germany, France etc.), and the share of these markets is expected to grow. It is not to be ignored that trade in live fingerlings is made from hatcheries in Italy, Spain and France to farms in Greece, Malta, Croatia and elsewhere.

As output of seabass and seabream has grown, market prices have more than halved between 1990 and 1999, dropping from US$16/kg to less than US$5/kg today. The rapid saturation of the market and the speed of the parallel decline in prices (down 50 per cent in five years, compared to the ten years for a similar drop in the case of Atlantic salmon) is attributed to several factors:

- The much smaller traditional market for these species (southern Europe), compared to Atlantic salmon;
- Lack of diversified products;
- Inadequate marketing and market development; and
- Absence of technological advances (*e.g.*, genetic improvement, efficient feeds and feeding strategies etc.) that could significantly improve productivity.

The substantial drop in prices should help to open new markets and expand existing ones, provided that acceptable profit margins can be sustained in production through improvements in productivity, diversification of products and intensified marketing efforts. American Catfish: American catfish is now the fifth most consumed fish species in the United States, measured at 0.4 kg/ capita edible weight in 1998. Exports are limited, as production targets the domestic market, but some exports have started to Europe. The reason for the success of catfish is similar to that of tilapia: consumer demand for white, easy-to-prepare fillets.

Seaweed

Farmed seaweed production has increased in the last decade (6.8 million mt in 1997) and is now 87 per cent of total seaweed supplies. Most of the output is used domestically for food, but there is growing international trade. China, the major producer, has started exporting seaweed as food to the Republic of Korea and Japan. The Republic of Korea, in turn, exports some quantities of Porphyra (red seaweed) and Undaria (brown seaweed) to Japan (1998 exports of 26 000 mt). Chile produces over 100 000 mt, and is responsible for some 75 per cent of global production of Gracilaria sp., which is used for agar manufacture. Significant quantities of Eucheuma (red seaweed) are exported by the Philippines, Tanzania and Indonesia to the United States, Denmark and Japan. Total EC imports of seaweed in 1998 amounted to 58,000 mt, with the Philippines, Chile, Indonesia and Australia as the biggest suppliers.

Molluscs

International trade in molluscs is limited when compared to total output, with less than 10 per cent of production traded. The major importing markets are Japan, the United States and France, while major exporters are China and the Republic of Korea. The contribution of these farmed products to trade is uncertain.

Oysters, followed by clams, scallops and mussels, lead farmed mollusc production, but international mollusc trade concentrates on scallops and clams (fresh and frozen). Total fresh and frozen scallop imports have grown from 28 000 mt in 1985 to 65 000 mt in 1998, reaching US$510 million. Clam imports have risen from 33 000 mt to 152 000 mt in 1997, valued at US$257 million. Mussel imports have shown a downward trend after a peak of 175 000 mt in 1992, but have bounced back to 188 000 mt worth US$220 million. Oyster imports have grown from below 10 000 mt in 1985 to 28 000 mt (US$112 million) in 1998.

Live Seafood

The cultural preferences and growing affluence in Asia indicate a clearly positive long-term trend of live seafood commerce. The live seafood market is largely restricted to the restaurant trade and to consumers with a relatively high disposable income.

Major market expansion is anticipated due to demand in China, but is also expected in Malaysia, Singapore and Taiwan Province of China, as well as in parts of North America with large Chinese communities. The potential for aquaculture to supply the market is promising. The sector is already supplying large amounts of shellfish and limited quantities of grouper, crabs and other species. Technological developments in the culture of preferred live food species will increase the contribution of aquaculture to supplies.

Seed Supplies

There appears to be significant regional and international trade in the seed of cultured aquatic organisms, mainly from aquaculture sources, but this is poorly documented at present in most instances. Mention has been made above of the regional trade of fingerlings for Mediterranean seabass and seabream, but there is also international trade in wild glass eels (*e.g.*, the recent development of large purchases of European eel elvers by China), "eyed" (fertilized) eggs of salmon and trout, the postlarvae of various cultured shrimps, Indian and Chinese carps, and others. There is also limited trade (in terms of quantity) in broodstock. Documentation of trade in seed and fingerlings will improve gradually as a response to concerns about the spread of pathogens and the movement of genetic material. One would anticipate that the concerns of traceability will also support such documentation.

7

Global Environmental Threats to Fisheries and Aquaculture

We live on a human-dominated planet, and the momentum of human population growth, together with the imperative for further economic development in most of the world, ensures that our dominance will increase.

SUSTAINABLE DEVELOPMENT AND MANAGEMENT OF AQUACULTURE AND FISHERIES SYSTEMS

The sustainable development and management of aquaculture and fisheries systems can only occur if these activities are well planned and integrated into the natural and social resource, ecosystems, and farming systems contexts of the larger global context of which they are a part. Population and natural resource constraints in a crowded future demand that aquaculture "fits" as a part of a larger strategy for the non-consumptive, multiple uses of water; and that fisheries be managed sustainably as part of the larger trends affecting the marine environment.

In the past 50 years there has been a massive migration—called the "greatest human migration of all time"—from rural, inland areas to the world's coasts, resulting in about 60 per cent of the Earth's people living within 100 km of the coast. Massive growth of coastal cities has raised issues of the future survival of coastal and estuarine ecosystems and habitats, and has put at risk the livelihoods of millions of people in traditional coastal communities that depend upon the sustainable capture and culture of aquatic living resources. Upwards of 35 per cent of the primary production of the temperate continental shelves is being harvested in fisheries, or discarded back to the sea as waste ("bycatch"). Coastal margins and oceans are among the most heavily used and modified areas of the planet, suffering disproportionate amounts of habitat destruction and pollution.

Humans have removed about 50 per cent of the world's mangroves. Human activities in the coastal zone deliver sewage, solid wastes, refuse (marine debris), sediments, dust, pesticides, and oil hydrocarbons to rivers, estuaries, and coastal areas. Assessing the range, magnitude and delivery of land-based sources of pollution to coastal oceans is a major global effort. It is estimated that about 80 per cent of marine pollution originates from land-based sources and activities.[2] Inputs of nutrients to the coastal zone from development and agriculture have caused an increase in toxic algae blooms, some of which have caused human disease and neurological impairment. In Shanghai, China, an estimated 13,000 factories discharged 10.55 million tons of waste in 1989, nearly twice the amount in 1980; about one-fifth was toxic waste. The Songhua River in China contains 10 tons of mercury and has waterborne mercury concentrations higher than those reported after the disaster in Minamata Bay, Japan. About one-third of China's coastal waters are polluted with oil; mercury and cadmium have been detected in 60 per cent and 33 per cent of seawater samples, respectively.[3] In Poland, 92 per cent of the nation's rivers are "beyond classification", meaning that the level of pollution is greater than that described by any existing pollution category.

Humanity is the now one of the major driving forces in the Earth's hydrological cycle, using more than half of the world's freshwater runoff. Agricultural water use accounts for about 75 per cent of total global consumption, mainly through crop irrigation, while industrial use accounts for about 20 per cent, and the remaining 5 per cent is used for domestic purposes. Most of the world's rivers are dammed—there are 36,000 dams—and the number of dams is increasing. Many major rivers (Colorado, Nile and Ganges) are so heavily used that little or no water reaches their deltas and the sea. Only 2 per cent of rivers in the USA run free. Major inland water areas (Aral Sea, Lake Chad, etc.) have nearly dried up or been greatly reduced to small lakes. It is estimated that two out of every three people will live in water-stressed areas by the year 2025. In Africa, it is estimated that 25 countries will be experiencing water stress (below 1,700 m^3 per capita per year) by 2025. About 450 million people in 29 countries suffer from water shortages.[5] Clean water supplies and sanitation are major problems in many parts of the world, with 20 per cent of the global population lacking access to safe drinking water. Water-borne diseases from fecal pollution of surface waters are a major cause of illness in developing countries. Polluted water is estimated to affect the health of 1.2 billion people, and contributes to the death of 15 million children annually.

Losses of biodiversity are occurring at alarming rates. Rates of loss are estimated at 100—1000 times greater than natural rates of extinctions. Extinction rates are even higher on remote islands, such as Hawai'i, USA. Due to the ease and rapidity of global transportation, the world's ecosystems are becoming "homogenized" due to "biological invasions": the exotic introductions of plants and animals, bacteria and other pests to natal ecosystems. After land transformations, exotic species impacts are the second major cause of species extinctions and population losses. About 11 per cent of birds, 18 per cent of mammals, 5 per cent of fish and 8 per cent of the Earth's plants are threatened with extinction, and exotic species have contributed greatly to their demise.

Given these immense global challenges, Vitousek *et al.* recommend:

- A slowing the rate of population growth since humanity's growth drives all resource use and waste generation,
- A reduction in the rate of human impacts since ecosystems can react to lower rates of impacts and stabilize with moderate levels of human-induced changes,
- Accelerating ecosystem level research and management understanding. Human impacts must be included in all analyses of global and ecosystem change. Ecological science must include development of new methodologies to assist humanity in this global crisis, requiring development of true interdisciplinary environmental scholarship, and education of the "informed generalist."

THE GLOBAL IMPORTANCE OF FISHERIES AND AQUACULTURE

Fisheries play an important role in the world food economy. Fisheries are a source of employment for about 200 million people who depend directly upon ocean fishing for their livelihoods. Fish is the primary source of protein for some 950 million people worldwide and represents an important part of the diet of many more. In less than 50 years, the world's average per capita consumption of fish has almost doubled. Globally, fish provide about 16 per cent of the animal protein consumed by humans, and are a valuable source of minerals and essential fatty acids. Fish is the primary source of omega-3 fatty acids in the human diet. Omega-3 fatty acids are critical nutrients for normal brain and eye development of infants, and have preventative roles in a number of human illnesses, such as cardiovascular disease, lupus, depression and other mental illnesses.

The reported production of fish for direct human consumption doubled between 1950 and 1970, and has stabilized since then at an average of 9.0 to 10 kg of fish per capita, notwithstanding world population growth. Fish consumption per person is expected to continue to rise. Supply will probably be limited by environmental factors, and a likely range for demand is 150 to 160 million tons, or between 19 and 20 kg per person in 2030. Global increases in consumption of food fish will take place predominantly in the developing countries, where population is growing and higher incomes are allowing purchase of high value fisheries items for the first time by many people.[11] However, fish production in least developed countries where fish protein is needed to prevent malnutrition is a key element of food security in these regions and a critical area where innovative programmes are needed to increase production.

Today, fishing is the largest extractive use of wildlife in the world. The value of world total fishery production in 1999 was US$ 125 billion. World production of fish, crustaceans, mollusks and plants reached 142 million tons in 2001. Capture fisheries production, which accounted for 66 per cent of the total, was 93.7 million tons, of which inland capture was 8.7 million tons, while aquaculture production was 48.4 million tons including plants. Marine and freshwater fish are also an increasingly important recreational resource, both for active users such as anglers and for passive users such as tourists, sports divers and nature-lovers.

PREDOMINATES IN CAPTURE FISHERIES AND AQUACULTURE PRODUCTION

Asia predominates in capture fisheries and aquaculture production, with China the leading nation worldwide. In 1999, over three-quarters (or 97 million tons) of the global production of fish, crustaceans and mollusks were utilized for direct human consumption. Fish utilized as raw material for the production of animal feed, *e.g.*, fish meal and fish oil, represented about one-quarter of the total fishery production in 1999; and this amount has remained relatively static for the past 15 years.[14] Most growth in the fisheries sectors is projected to occur in developing countries, which will account for 79 per cent of food fish production in 2020.

Increased production of fish from aquaculture has occurred primarily as a result of increasing feed inputs into ponds and other production systems, thereby increasing yields per hectare by an order of magnitude compared to extensive production systems in which rearing water is fertilized only. Higher inputs mean two things to the aquaculture feed industry; more feed and higher quality feed. Currently, global feed production for farmed fish and crustaceans is approximately 13 million tons, and predictions are for feed production to increase to over 37 million tons by the end of the decade, an increase of 24 million tons. Feeds for salmonids and marine fish have always been complete feeds, *i.e.*, ones that supply all of the nutritional needs of the fish. Pond-reared fish, in contrast, obtain a significant proportion of their nutritional needs from pond biota. The degree to which feeds must supply essential nutrients to pond-reared fish increases as rearing densities increase beyond the capacity of natural foods in ponds to supply them. Fish farmers around the world have found that as they increase feed inputs, the biomass and economic yields from ponds increase as well. Thus, great areas of low-input, pond-based aquaculture mainly in Southeast Asia and China are being converted from low-input systems to high-input systems that depend upon high quality feeds to supply an increasing proportion of nutrients used by the fish.

The effects of the aquaculture industry's growth and of changes in feed input in pond-based aquaculture have been dramatic with respect to the use of marine proteins by the aquaculture feed industry. In the mid-1980s, less than 10 per cent of annual fish meal production was used by the aquaculture feed sector. Today, that proportion is over 40 per cent.[16] Similarly, aquaculture now uses nearly 75 per cent of annual global fish oil production, up from less than 10 per cent seventeen years ago, but this change is mainly due to the adoption of high lipid feeds by the salmon farming industry, rather than increasing the inputs into pond-based aquaculture systems. Fish meal and oil are produced from species of fish that are not generally utilized directly for human food, *e.g.*, herring and capelin in Norway and Iceland, sand eel in Denmark, capelin in South Africa, anchovies in Peru and northern Chile, jack mackerel in central Chile, sardines in Japan, and menhaden in the USA. The rapid increase in fish meal and oil use in aquaculture feeds over the past 15 years has not resulted in over-exploitation of the stocks of fish harvested primarily to produce fish meal and oil. Production of fish meal has averaged between 6-7 million tons per year since the mid-1980s, except in El Nino years, when production has been lower. Fish oil production has averaged between 1.2-1.3 million tons. No significant changes in annual harvest or fish meal/oil production associated with increased aquaculture production or increasing intensification of aquaculture and concomitant higher feed inputs is evident. Increased fish meal/oil use in the aquaculture sector has come at the expense of other uses.

Estimates of future protein requirements for aquaculture feeds depend upon future production from various segments of the aquaculture industry and on annual production levels of fish meal. As aquaculture production expands, it will be essential to replace portions of fish meal and oil in fish feed formulations with alternative ingredients derived from alternative sources, primarily grains and oilseed products.

Because 70 per cent of the fish meal used in diets for fish and crustaceans is used to produce diets for salmonids, marine fish and shrimp, these production sectors are the focus of most research with respect to the use of alternative protein sources. Numerous studies have been conducted to evaluate the effects of replacing various percentages of fish meal in diets for these fish, and, without exception, none has successfully replaced 100 per cent of fish meal without reducing fish performance. At best, 50 per cent of fish meal in diets for salmon and trout can be replaced by soy protein concentrate, and 25-30 per cent with soybean meal. Similar findings have been reported in studies of wheat gluten meal, corn gluten meal, and rapeseed protein concentrate. For the most part, formulated fish diets are used to produce high-value fish for export. In underdeveloped countries, increased production of lower-value fish that are consumed by local populations will depend upon the use locally-available, low-cost feed ingredients that can be combined to produce prepared feeds to increase productivity of community ponds and waterways.

This is an area that requires substantial investigation and development, and is a logical target for USAID support. Another logical target for USAID efforts is the development and testing of feeds that are based upon sustainable plant-derived protein sources for use in developing countries where fish are grown both for domestic consumption and export, such as China. Partnerships with US commodity groups could be a mechanism to extend USAID efforts. It is critical that USAID become a leader in this area, given the perception that increased aquaculture production leads to higher fish meal use and greater pressure on stocks of fish that are captured to produce fish meals and oils.

Fisheries products have become the most internationally traded food, as some 37 per cent (by quantity) of all fish for human consumption is traded across borders. In 1999, international trade (in live weight equivalent) represented 34 per cent of the total production. In 1999, foreign trade earnings amounted to US$ 52.9 billion. Most fishery exports were destined to developed countries, with Japan the world's leading importer, and the USA second. The US balance of trade deficit in seafood products is approximately $9 billion.

Exports of fisheries products to the developed countries have become so lucrative that nations like Argentina, a traditional, globally important exporter of meat products and livestock have turned to seafood exports as the major source of foreign exchange earnings. In 2002, exports of raw and processed fish and shellfish from Argentina surpassed meat products and livestock, with beef earning US$ 574 million in export revenue, while seafood—mostly prepared products—earned US$ 714 million. Argentina's main consumers of domestic seafood products were Spain, Brazil and the US.

ENVIRONMENTAL IMPACTS OF FISHERIES

In the process of producing food, economic resources, employment, livelihoods and recreation, fisheries modify ecosystems through more or less reversible impacts on the target resource itself (including through overfishing), the species associated to or dependent on it (such as predators or preys), on the trophic relationships within the ecosystem in which the fishery operates, and on the habitat. The fishery sector is expected to reduce these impacts to a minimum that would also be compatible with its own existence.

Excessive fishing reduces spawning biomass below optimum levels and catches below the maximum sustainable yield. The problem, systematically addressed for a century albeit with mixed results, began in the North Atlantic before World War II and spread progressively to the North Pacific in the 1950s, the Eastern Atlantic and Eastern Pacific in the 1960s, the Indian and Antarctic oceans in the 1970s and the South Pacific and Southwest Atlantic in the 1980s - all with very significant economic costs.

Changes in species composition and biodiversity are common with progressive reduction of large, long-lived, and high-value predator species and the increase in small, short-lived, and lower value pelagic and demersal prey species. The process has been coined as "fishing down the food chain".

Important macroscopic changes have been observed in many ecosystems such as the North Sea, Yellow Sea, North Atlantic (*e.g.*, George's Bank and

Barents Sea), Gulf of Thailand, and Southeast Australia. In some regions, the abundance of scavenging birds has significantly increased, possibly due to the large quantities of discards. It is now also clear that intensive fishing reduces genetic diversity of wild populations (*e.g.*, rapidly depressing the proportion of fast growing and late spawning individuals). Changes in species composition or dominance can also be provoked through competition for food between fisheries and marine apical predators.

NON-SELECTIVE FISHING

Non-selective fishing gear Use of non-selective fishing gear (*e.g.*, trawls, longlines, gillnets) can take a significant bycatch of juvenile fish, benthic animals, marine mammals, marine birds, vulnerable or endangered species, etc. that are often discarded dead (about 20 million tonnes per year). While bycatch and discard problems are usually measured in the potential loss of human food, the increased risk for particularly vulnerable or endangered species (*e.g.*, small cetaceans, turtles) is significant. In the North Sea, for example, the impact on the food chain and species composition is consequential as discards can represent up to 30 per cent of birds' consumption.

Ghost-fishing Ghost fishing occurs when certain gear such as pots or gillnets continue to kill fish for some time after having been discarded at sea or lost, *e.g.*, in bad weather.

Impacts on the bottom The intense use of trawls and other mobile bottom gear (*e.g.*, dredges) result in changes in bottom structure, microhabitats, and benthic fauna. The effect is particularly obvious when these gears are used in environments such as sea grass and algal beds, coral reefs, sponges, tube worms. The longterm impact is less obvious on soft bottoms although scraping and ploughing the bottom to depths of as much as 30 cm can seriously disturb the substratum habitat and productivity.

Destructive fishing practice Destructive fishing practices have also spread in some poor coastal communities (*e.g.*, the use of dynamite and poison) - particularly on coral reefs. Other less conspicuous or debated impacts relate to direct dumping of debris (gear, twine, food containers, plastic bands, etc.) by fishing fleets, including organic pollution by at-sea processing factory ships, coastal pollution of processing plants, contribution to global warming through exhaust fumes and refrigerant gases, and accidental introduction of alien species.

Solution

Overfishing is the central problem in fisheries management and must be tackled through better control of fishing capacity and specific resource rebuilding plans. The same set of measures would also improve species composition although it is impossible to exploit the ecosystem without affecting the relative importance of different species.

Selectivity of gear can be improved and impacts on the bottom and habitat can be reduced through better gear technology and improved fishing practices. Ghost fishing can be decreased through greater awareness raising, prohibition and control of dumping of damaged gear at sea, as well as active at-sea programmes for retrieval of lost gear. Gear technology can also contribute in diminishing the fishing capacity of lost gear (*e.g.*, biodegradable material, collapsible traps, etc.). Destructive fishing practices must be further prohibited with stronger enforcement. Overall, however, the solution is also in generally improving the economic conditions of poor coastal communities and providing alternative forms of livelihood. In general, substantial efforts are needed to improve data collection and research on these issues.

Action

Countries have been tackling overfishing for a long time, although not very successfully. On the institutional side, NGOs have become very active at uncovering and illustrating the problem, exerting considerable pressure on governments and international fishery organizations. An International Plan of Action (IPOA) for the Management of Fishing Capacity was adopted at FAO in 1999 and is currently being implemented. In addition, the FAO IPOA on Illegal, Unreported and Unregulated Fishing (IUU), adopted in 2001 will also contribute to countering overfishing. In order to limit international trade in endangered fishery species, CITES is considering adding fish species subjected to large-scale exploitation to its annexes. The progressive development of ecolabelling practices and the increasing integration of environmental criteria into sustainability criteria for fisheries should also provide additional incentives and fuel consumers' pressure.

On the operational side, much has been done to address the environmental impacts of fisheries. Some countries have regulated discarding, imposing severe quotas, or banning it altogether, forcing the landing of all unwanted bycatch (*e.g.*, in Norway, Canada, Iceland, and the Faeroe Islands). Selective grids, panels and square meshes are used in several trawl fisheries to facilitate the escape of unwanted species or small-sized individuals. Longlines are being modified to reduce bycatch of birds and an IPOA for Reducing Incidental Catch of Seabirds in Longline Fisheries was adopted by FAO in 1999.

Special devices are currently used in tuna fisheries to successfully reduce dolphin catches. Zoning has been conventionally used to keep trawlers away from vulnerable coastal habitats, although with little success in areas with large overcapacity. Marine Protected Areas (MPAs) have become fashionable and if properly enforced could be more effective. In some countries (*e.g.*, Norway), programmes exist to retrieve lost gillnets lying on the bottom. The development of integrated and more sustainable livelihoods programmes are being implemented (*e.g.*, by FAO in Western and Central Africa). Some countries

(USA, Ireland) require the elaboration of an environmental impact assessment (EIA) and/or environmental impact statement (EIS) for their fisheries.

Outlook

Awareness of this issue has been greatly increasing since UNCED both in the fisheries sector and among the public. The general pressure and progressive evolution towards ecosystem-based fisheries management will facilitate the reduction of environmental impacts. The sector is quite willing to act positively if this can be done in an economically viable way. Progress made in fishing and processing technology is significant and the position of the fisheries sector regarding ecolabelling is gradually improving. As a consequence, while much remains to be accomplished, the prospects for improvement are optimistic.

DESTRUCTIVE FISHING PRACTICES

All fishing methods have an impact on the target resource and may also affect non-target species. Many of them also have an impact on the wider aquatic environment. But the "normal" effect of exploitation should not be confused with "destruction" which, according to the dictionary implies "to reduce to a useless form, to spoil completely; to put out of existence; to obliterate, wipe out, annihilate, demolish, devastate, tear down, raze". The term "destructive fishing" has been generally used by scientists and environmental NGOs for a wide range of situations spanning from classical overfishing (excessive use) to outright destruction of the resource and its environment with explosives (wrong methods). Indeed, many fishing gears could be considered "destructive" if used in the wrong environment.

In a fisheries management perspective, it is however important to distinguish among:

Overuse or overfishing, to be corrected mainly through fishing capacity reduction; non-selectivity, which in most cases can be corrected through improved gear technology, closed seasons, or closed areas, etc.; destructive fishing, or destructive use, *e.g.*, when a gear is used in the wrong habitat, such as bottom trawls in seagrass beds where its impact is unacceptable, and where the gear can be prohibited; and destructive methods, the impact of which are so indiscriminate and/or irreversible that they are universally considered "destructive" in whatever circumstances. Among the latter, the use of poisons and explosives is prominently cited in the literature and the muroami fishing practice is often referred to (in Southeast Asia). The economic losses for the local and wider society resulting from such damage far outweigh the short-term individual gains made by the users of these destructive methods.

Poisons The use of poisons is widespread, in some regions in both fresh and marine waters - especially in coral reefs and coastal lagoon fisheries. In many places the use of poison is a traditional practice but the effects have been exacerbated by the use of pesticides to replace poisons of vegetable origin. As fish become scarcer through overfishing or in order to catch rare, small and precious aquarium fish, local fishers often resort to using poisons such as cyanide or pesticides. Fishers can easily obtain inexpensive cyanide used in the jewellery industry and gold mining. Pesticides are readily available to farmers, which are often also part-time fishers. Techniques used vary across regions/localities. They are effective at killing or stunning, indiscriminately, the fish, which are then collected by divers, or through netting and seining. The poisons kill also other organisms from the ecosystem, including the coral reef-building organisms.

Explosives Fishing with explosives, also known as blast fishing, has probably been in existence for centuries and is apparently spreading. Explosions can produce fairly large craters, devastating 10-20 square meters of bottom. In coral reefs, recolonization of damaged habitats is very slow and complete recovery may take several decades. The explosion kills both the target fish and the accompanying fauna being indiscriminate in size or species. In many instances, errors of manipulation have lead to injuries and death of humans. Explosives and the raw materials for preparing explosives such as fertilizers and sugar are cheap and easily available. Commercial explosives are often obtained from mining or construction activities. In many areas, fishers only need to extract the explosive charges from munitions left over from on-going or past armed conflicts. In other areas, fishers can access army munitions through illegal channels. Explosives can have very serious consequences for the resources, the environment, and unfortunately sometimes also for the users themselves.

Muroami The Muroami fishing technique, employed on coral reefs in Southeast Asia, uses an encircling net together with pounding devices. These devices usually comprise large stones fitted on ropes that are pounded onto the coral reefs. They can also consist of large heavy blocks of cement that are suspended above the sea by a crane fitted to the vessel. The pounding devices are repeatedly and violently lowered into the area encircled by the net, literally smashing the coral in that area into small fragments in order to scare the fish out of their coral refuges. The "crushing" effect of the pounding process on the coral heads has been described as having longlasting and practically totally destructive effects. Inadequate practices Some standard fishing gears could be used in a way which damages the resource and/or the environment, to such an extent that they could be (and have been) considered as "destructive" fishing practices. These may include, inter alia, the following practices that can be properly regulated and controlled:

- Beach seining, because of the large proportion of juveniles yielded and often discarded;
- Bottom trawling because of its impact on the bottom and on bottom-dwelling animals and benthos such as sea stars, urchins, clams, etc.

In addition, when improperly used in the wrong environment *e.g.*, on coral reef areas or coastal seagrass beds (despite of being generally prohibited in such areas) it may have very longlasting effects on habitat;

- Large-scale pelagic driftnets because of their ability to ensnare large marine animals such as mammals, sharks, turtles, and a number of vulnerable species;
- Fishing with respiratory assistance using scuba equipment or compressors, because their effectiveness is judged excessive; anchoring of vessels in reef areas (particularly the Staghorn coral) which breaks them, seriously damaging the whole reef sustainability.

Solution

The techniques universally recognized as "destructive" can only be explicitly banned and their use severely punished. Most bad ways of using fishing gear have been identified and are prohibited in national legislation. However, experience shows that this is not sufficient because in stressing economic situations, the incentives to misbehave might be very high.

For the poor, poisons and explosives are among the cheapest method available, requiring little capital investment. In some instances, it is the only affordable access to a vital resource. Often, however, the replacement of poison and explosives by other types of fishing methods is within the financial means and reach of those who currently use these destructive fishing practices. For the poorest, however, it will be difficult to suppress these bad fishing habits without improving their economic condition. In these instances, the long-term solution to the use of destructive methods is in improving livelihoods or providing affordable alternatives.

The Muroami technique does not seem to be associated specifically with poverty. It reflects a total disregard for sustainability and as such should be banned.

The destructive modes of use of other more standard gears can also be "promoted" by chronic overfishing, in a very damaging vicious circle. Proper practices should be regulated through protection of particularly important areas, or habitats (seagrass beds, coral reefs) as well as closed seasons (particularly to protect juveniles or sensitive species).

Eliminating destructive methods requires finding solutions to a number of issues affecting the fisheries sector such as overfishing (an example of "the hen and the egg" dilemma), inequitable resource allocations, food insecurity and poverty. In most cases, there is a need to improve stewardship over the resources through better monitoring and control, educating fishers about the destructive nature of their practices, and the establishment of explicit forms of use rights for local communities together with decentralised management responsibilities as well as no-take areas. There is evidence that the allocation of use rights will create a strong incentive for communities to use responsible fishing practices and persecute and punish trespassers.

Action

Awareness about these methods and their effect is growing. All national fishery laws generally prohibit poisons and explosives but enforcement is inadequate. Large-scale pelagic driftnets have been banned by the United Nations General Assembly. The use of beach seines is prohibited in many countries and areas. The coastal areas with coral reefs or coastal seagrass beds are usually prohibited to trawls.

It should be recognized, however, that enforcement is often imperfect, particularly when offshore resources are overexploited. Programmes have been implemented to educate small-scale fishing communities of the damage caused by explosives and cyanides. The devolution of fishing rights and responsibilities to coastal communities (*e.g.*, in Philippines) and their greater involvement with other stakeholders and NGOs in the management of resources, accompanied by programmes to raise awareness of the long-term damage of such fishing techniques, have contributed to the promotion of local social pressure and enforcement to control destructive fishing. In addition, the development of tourism can increase the value of the fish and the environment creating incentives for their better conservation, provided the economic benefits of development are equitably distributed.

Outlook

More of the actions mentioned above are still needed to reduce destructive fishing. They include measures aimed at suppressing overfishing, improving livelihoods, and reducing poverty that all contribute to reducing the probability of and need for using these methods. The evolution of the problem in the future largely depends on local conditions and it is difficult to generalize.

At the moment, and with the little information available, the problem seems to be spreading. Once described mainly in Asia and Africa, it is now becoming more common in the Mediterranean and Latin America. It is unlikely to regress without the widespread establishment of explicit use rights in fisheries. Moreover, the complete cessation of such destructive practices is likely to require that the economic situation of the poorest strata of the fishery and rural sectors improves significantly.

EMERGENCIES IN MARINE FISHERIES

Emergencies in marine fisheries and aquaculture are triggered by a variety of events, including climatic and environmental emergencies, and disease outbreaks. The most important measures that can be taken to minimize the impact of emergencies are of a precautionary nature. By anticipating threats, making contingency plans and by taking appropriate precautions, much damage, which might otherwise occur in an emergency, can be avoided.

The frequency and intensity of storms have been increasing during the past decade. Fortunately, marine organisms and their habitats tend not to be seriously affected directly by storms. However, fishers face the risk of losing their lives in storms and suffering the destruction of their capital, such as boats and fishing gear.

An important possible measure might be installing or improving storm-warning systems, involving shore-to-boat and boat-to-boat communications networks. Other measures that might help mitigate the impact of storms are the training of fishers in safety precautions.

Environmental emergencies may occur as a result of oil spills, releases of hazardous chemicals, or the introduction of harmful organisms (*e.g.*, toxic algae blooms). These can threaten the survival, health and reproduction of fish stocks, whether wild or farmed. Disease outbreaks at epidemic levels in the aquaculture sector often constitute an emergency. Contingency planning to deal with such outbreaks should aim to reduce the consequent spread of disease as early action can significantly lessen the social and economic impacts that otherwise can result.

Fisheries insurance schemes offer another means of dealing with unavoidable damage caused in emergencies. Although there is growing use of such schemes, there is still significant scope for promoting them further, particularly in developing countries.

CLIMATIC EMERGENCIES

Storm-related disasters have been increasing in frequency and intensity during the past decade. The high winds that accompany tropical storms and the resulting floods have a particularly devastating impact on food production. Although on average, hurricane intensity has remained essentially steady for the last three decades, there is some evidence to indicate that their frequency may be on the increase. Further, the devastation caused by tropical storm disasters has been rising enormously in the 1990s, owing in part to the increase of population in storm-prone areas.

The most recent World Disaster Report of the International Federation of the Red Cross shows that during the ten-year period 1990-1999 wind storms and flood related disasters combined accounted for 60 percent of the total economic loss caused by natural disasters. A significant share of disaster casualties, in human suffering in terms of lives lost and injured as well as people displaced from their homes and livelihood activities is also attributable to storms and floods. The major cause of the most destructive phenomenon is the storm surge - a rapid rise of the sea level resulting from strong winds driving the water ashore and causing flooding in low lying coastal areas. In low lying coastal areas (*e.g.*, highly fertile and densely populated river deltas), storm surges account for more than 90 per cent of loss of life and property.

Outlook

Despite increased flood awareness and cyclone warning measures, the last 10 years have seen a 300 percent rise in the number of individuals affected by floods and storms. Between 1973 and 1997 hurricanes, cyclones, typhoons, storms and tornadoes claimed, on average, each year an estimated 11 000 lives and made more than more 1.1 million people homeless. In Bangladesh alone three storms, four floods, one tsunami and two cyclones killed more than 400 000 people and affected another 42 million during this period.

The World Watch Institute estimated that during the first 11 months of 1998, weather-related disasters caused more than $89 billion in economic losses (as compared to $55 billion during the 1980s), resulted in 32 000 deaths and displaced 300 million people from their homes and livelihood systems. Most of the disasters in 1998, including record floods in China and Bangladesh and Hurricane Mitch in Central America, were attributed to the El Niño or La Niña phenomena.

Although no aggregate quantitative estimates are available the economic cost of storm-related disasters in 1999 and 2000 was also considerable. For instance, the direct and indirect economic cost of the floods in Mozambique caused by Tropical Storms Elyne and Gloria in February and March 2000 is estimated at US$1 billion, as compared to the country's export earning of only US$300 million in 1999.

Impact of Storms on Fisheries

While some concern may be expressed at the damage caused by "ghost fishing" (where lost fishing gear continues to catch fish), marine life and their habitats are not seriously affected by storms. If boats, fishing gear and associated essential infrastructures have not been lost, fish production can resume immediately following passage of the storm.

The most serious impact of storms in the fishery sector is risk faced by fishers in terms of their lives and/or destruction of their capital such as boats and fishing gear. For example, in November 1996, following a severe cyclone, approximately 1,435 fisherfolk were reported dead or missing in the State of Andra Pradesh on the east coast of India. Of these, 569 were reported to have been lost at sea while fishing in mechanised boats, while 830 fisher folk were lost while carrying out shrimp seed collection and other shore-based activities in areas remote from their villages. The causes of death in these two activities are of a totally distinct nature, whereas the former were lost at sea in conditions of high winds and heavy seas, the latter died on land, largely as a result of the storm surge.

Storm-related damages to capital assets, including boats and fishing gear, means the loss of income and livelihood, especially for poor fishing households. Poor households who depend on fishing for their livelihood may not have adequate saving to replace their capital, and are likely to face the risk of food insecurity in the aftermath of storms unless their capital is replaced immediately through public support. The disruption of fishing activities by such households could also affect the livelihood and food security of other households, for instance, small traders who buy and sell fish in small local retail markets.

DISEASE OUTBREAKS IN AQUACULTUE

When a disease breaks out at epidemic levels in an aquaculture sector, it frequently constitutes an emergency. Such emergencies can happen at all levels; farm, local area, national, or regional. The most important consideration for avoiding such emergencies is proper advance planning and preparedness.

Solution

Advance planning to deal with serious disease outbreaks can significantly reduce the social and economic impacts that result. In addition, prompt action based on solid contingency planning can reduce the consequent spread of disease agents. Using a fire outbreak as an analogy, economic losses will be lower if the fire is detected quickly and fire fighters arrive at the site promptly, and with the resources necessary to stop the fire. The speed of their arrival on the scene depends on the efficiency of the reporting system, the speed of their response and the availability of fire fighting equipment. The effectiveness of the fire-fighting team depends on their training and their experience under different conditions. Preparation and response to aquaculture disease can be looked at in a similar way. To minimise losses, it is necessary to have good surveillance, accurate disease diagnosis, efficient reporting systems and well-trained specialists who know how to deal with a variety of potential disease emergencies. The needed response may be the same at government/ institutional level and at the farm level, but the extent and manner of action may differ between the two.

It is difficult to predict when a new disease might appear, or how much of an impact it is likely to have on a particular aquatic resource. Known diseases with established etiological information are more predictable and thus more easily dealt with. Some disease outbreaks are small and/or localised, and so are more easily controlled. Others may be large, spread rapidly and be difficult to manage. Disease outbreaks are influenced by different factors, including weather, geographic isolation, and transmission dynamics. All these factors can affect the ability of personnel to respond to, contain and tackle a disease outbreak. Private companies should be included in the overall planning process of how disease outbreaks should be dealt with, in close partnership with local, state, and central government agencies. Strategies can be developed based on the information available, including: what diseases are known to occur within an area, and how frequently; conditions which are known to predispose the animals (or plants) to disease; the proximity to other farms and areas where the disease may occur; and, knowledge of any extreme weather conditions that might occur in the area at different times of year.

Contingency plans should be designed to prepare for the kind of disease outbreak that is "most likely" to occur in a particular area or facility. On rare occasions a new disease occurs, or the impact of a disease is greater than anticipated. To prepare for these unusual but significant incidents, contingency plans must also include "worst-case" scenarios, such as for a highly infectious disease that may spread rapidly and cause heavy mortalities.

One difficulty with "new" disease situations is deciding at which point it is serious enough to warrant classification as an emergency. Actual diagnosis of the disease may be of limited value in the decision making process, since the origin and cause of the problem may be unknown. Samples should be sent for analysis to obtain pathology information, but the time required for laboratory processing is often too long for the results to be of immediate assistance to a farmer suffering acute or severe mortalities of his stock. This is particularly so when the analysis requires identification of a new pathogen and investigation of its modes of transmission. However, firm diagnosis of the disease is not always necessary for decisions to be made on interim control measures. Many diseases have been first described in the past only on the basis of their gross pathology or characteristic features. These descriptions allow farmers or extension staff to make a presumptive diagnosis with a clear, consistent case definition and reach an early a decision on the best disease control measures. Laboratory results may then reinforce or refute the presumptive diagnosis, increasing the level of diagnostic certainty and permitting the refinement of effective control strategies.

Action

A carefully designed contingency plan will describe the major actions to be undertaken when a disease occurs. To optimize efficacy and minimize the spread of a disease, these actions should take place immediately following detection or reporting of the outbreak and should include:

Notification of all staff, individuals, private companies or government agencies that are responsible for the disease control effort; notification of all those likely to be affected by the disease or the proposed control measures; mobilizing trained personnel to the site quickly; determination of the extent of the disease, its nature, speed of transmission, and likelihood of spread to neighbouring farms, sites or the local environment; taking measures to prevent any further entry of the disease agent onto a site or into a population; confining the outbreak to a limited area; eradicating the disease where possible (usually this is only feasible in land-based facilities or isolated ponds - there have been few cases of successful eradication of a disease agent from open-water populations); rapid removal of moribund or dead animals from the water, followed by sterile or landfill disposal; follow-up surveillance and monitoring after the disease outbreak is brought under control.

Outlook

At the farm level, the complexity, but by no means the efficacy of a contingency plan, depends upon the size and scale of the operations. Small farms can usually manage with basic plans such as gross surveillance and monitoring, due to their straightforward organizational structure (owner-operated, owner and manager, owner and technician) where everyone has clear roles and responsibilities and good lines of communication exist. Larger farms often have a more complex division of responsibility and thus need more detailed contingency plans. The ability of farms and organisations to co-operate in dealing with disease emergencies depends partly upon the awareness of the short and long term benefits of good planning.

Cooperation is best served by a clear assignment of responsibility to the different staff members or to the various representatives of the farms and organizations involved.

ENVIRONMENTAL MANAGEMENT SYSTEMS

Environmental management systems are also starting to evolve for some forms of aquaculture, as well as more formal environmental management systems (EMS). EMS is a complex approach with little direct application to small-and medium-sized farms, however, EMS principles provide useful guidance for improving environmental management of aquaculture production systems (H. Dixon, pers. comm.).

EMS assemble management policies, programmes and practices designed to identify links between industry, urban and developmental activities, and consequent pressures on the environment. An effective EMS for aquaculture should establish indicators of changes in the environment, including land, water and aquatic resources. Policies and practices responding to the changes are implemented with continuous feedback to reduce/mitigate any environmental impacts. These indicators should make clear links between environmental impacts and aquaculture activities.

The indicators should also reflect positive and negative impacts of environmental influences on aquaculture, as well as visa versa. EMS provides a possible systematic approach to motivate aquaculture to better organize priorities and projects to identify problems and potential impacts before they occur, as well as meet environmental and business goals. This process also assists compliance with national environmental laws and regulations. A successful EMS provides the means by which aquaculture can identify causes of environmental problems and prevent them, thus saving money to repair or mitigate after the damage has been done.

According to H. Dixon, key benefits of an EMS include:

- Improved environmental performance,
- Reduced liability,
- Competitive advantages,
- Improved compliance,
- Reduced operational costs,
- Enhanced consumer trust, and
- Increased access to capital.

H. Dixon considers that formulating an institutional framework for environmental management systems of aquaculture should:

- Be simple to implement, clear and comprehensible to all involved, including the public;
- Consider the needs of all stakeholders;
- Consider individual abilities and resources;
- Be financially flexible and not inhibit the activities it is designed to address; and
- Be based on sound scientific information, quantifiable and effective.

CODES OF PRACTICE

The technical methods, management systems and practices needed for minimizing impacts are being increasingly incorporated into more formal "Codes of Practice", notably in more commercially oriented and intensive shrimp and salmon farming. Several aquaculture organizations, for example, the Marine Shrimp Culture Association of Thailand, the Global Aquaculture Alliance, the Australian Prawn Producers Association, the Irish Salmon Growers Associations and others have taken the FAO Code of Conduct a step further and formulated Codes of Practice (COP). These COP contain principles for preventing or mitigating negative environmental and social impacts through use of "best management practices" (BMP).

They are currently for voluntary adoption and consist of documented guidelines available to farmers. Much needs to be done towards their implementation, such as the development of operational manuals and support programmes. Furthermore, the extent to which COP will be fully adopted by farmers under self-regulation and the environmental consequences of their adoption remain to be determined. Implementation is a very important issue for the new millennium, particularly for small-scale farmers.

Improving Profitability of Aquaculture Operations

The systems approach is directly applicable to the business of aquaculture, through the use of a structured, systematic approach to operations and business management.

E. Hempel considers that in high technology salmonid aquaculture, the "software" of aquaculture (people) is more important than the hardware or "technology". There are many facts and variables, so in order to understand and manage the system, things have to be simplified-"systematized"-. The approach emphasised by E. Hempel involves:

- Understanding and identifying the system of concern within the farm enterprise/business;
- Simplifying the system by identifying only key system variables;
- For each person working at different levels of the organization, identifying the key system variables to monitor;
- Reducing the number of variables to the minimum-*e.g.*, dissolved oxygen, food conversion ratio (FCR), cash flow, and set limits/ standards; and
- Monitoring and responding to changes that exceed the set limits.

E. Hempel also emphasises that effective management based on a systems approach requires assigning of responsibilities within an aquaculture business to monitor and respond to changes that occur. He also emphasises that the system may change with time, and that management has to be flexible to respond to such changes. He further considers that the systems approach to business and management is important to improving the performance and profitability of farming, and is relevant at all levels of aquaculture, from small-scale farms to the largest business.

Institutional use of a Systems Approach

A systems approach can also be used to define institutional responsibilities in aquaculture. For example, in western Australia, according to C. P. Rogers' presentation during the "Systems session" of the Conference on Aquaculture in the Third Millennium, NACA/FAO, 2000) a systematic analysis of institutional responsibilities was useful to establish the key points for decision making in allocation of land and water resources for aquaculture development. This helped streamline permit processing. He also emphasised that a systematic evaluation of management processes can be valuable in identifying institutional responsibilities at both government and private-sector levels, as well as for promotion of small-scale aquaculture.

Future Aquaculture Development

Aquaculture can be socially, environmentally and economically sustainable and, contribute to the production of food and rural development, provided appropriate farming systems and management practices are adopted.

The multidisciplinary and multisectoral systems approach recognizes that technical, economic, social and environmental issues, as well as institutional factors, have to be considered in the process of development and management of aquaculture. The systems approach attempts to understand the way the system operates and the interactions between different components, and serve as a basis for better management. This multidisciplinary approach requires different skills and, as such, also needs cooperation among different disciplines and information exchange among different stakeholders. The application of a systems approach to aquaculture has several implications for a more analytical and structured approach to aquaculture development.

There is a need for better information on aquaculture systems and promotion of more effective information exchange between stakeholders. Information requirements include social and environmental interactions at the farm level, development of better practices targeting important environmental and social impacts, and seeking incentives for farmers to adopt better farming practices. Development of suitable indicators is also another important factor. For assessing natural resource use by aquaculture, recent research has emphasised the "ecological footprint". The aims of the authors to develop a "sustainability" index are to be commended, however, there are a number of problems with the models promoted so far, including:

- Their static and dimensional (area) nature;
- The difficulties of incorporating economic and social values into the approach;
- The emphasis on biophysical factors;
- The lack of the footprint and ecological capacity to accept multisectoral or multiple uses;
- The lack of inclusion of ecosystem services (such as absorption capacity for discharge of nutrients into coastal ecosystems); and
- Difficulties in making comparisons between different locations and systems.

There was a consensus in the Conference discussions on the need to develop and evaluate meaningful indicators of resource use efficiency to better define environmental interactions with aquaculture and improve management decisions. Research is required to develop and evaluate meaningful indicators of resource use efficiency useful to management decisions.

Institutional strengthening in the private and public sectors is another critical issue that must be addressed.

The systems approach recognizes that national institutions (government and non-governmental agencies) are the primary facilitators of aquaculture development in their countries. However, national institutions involved in the promotion of aquaculture need to increase their capacity and reinforce emphasis on multidisciplinary approaches. This will require a shift away from traditional "top down" and "technology focused" capacity building and extension approaches, towards an approach that is interactive and responsive to farmer needs.

As emphasised above, much aquaculture research to date has focused on technical questions, leaving the impression that the major constraint facing aquaculture is a shortage of technical knowledge. This detracts from the attention needed for development and basic aquaculture education. Another important constraint to aquaculture is dissemination of existing knowledge, whether derived from research or the indigenous technical knowledge of farmers. The limited capacity of many developing countries' national institutions in education, research and development concerning the promotion of aquaculture compound this problem, and require serious attention in the new millennium.

Examples of the systems approach mentioned during the panel discussions included:

- Its application to small-scale farmer research and extension;
- The development and application of best management practices to improve environmental management on commercial aquaculture farms;
- Improving the profitability of aquaculture operations through the use of a structured, systematic approach to operations and business management; and
- The institutional use of a systems approach to assist in effective decision making at a regulatory/planning level and within the aquaculture industry as a whole.

SUSTAINABLE AQUACULTURE

INTRODUCTION

Aquaculture is currently one of the fastest growing food production systems in the world. Most of the global aquaculture output is produced in developing countries and significantly in low-income food-deficit countries. As defined by the united food nations Food and Agriculture Organisation (FAO), aquaculture is the "farming of aquatic organisms including fish, mollusks, crustaceans and aquatic plants. With stagnating yields from many capture fisheries and increasing demand for fish and fishery products, expectations for aquaculture to increase its contribution to the world's production of aquatic food are very high, and there is also hope that aquaculture will continue to strengthen its role in contributing to food security and poverty alleviation in many developing countries.

However, it is also recognised that aquaculture encompasses a very wide range of different aquatic farming practices with regard to species (including seaweeds, molluscs, crustaceans, fish and other aquatic species groups), environments and systems utilised, with very distinct resource use patterns involved, offering a wide range of options for diversification of avenues for enhanced food production and income generation in many rural and peri-urban areas.

WHAT IS SUSTAINABLE DEVELOPMENT

Though living resources are self-renewable, they have to be utilised rationally on a sustainable basis in harmony with the environment. Sustainable development is the management and conservation of the natural resource base and the orientation of technological and institutional change in such a manner as to ensure the attainment and continued satisfaction of human needs for present and future generations. Such sustainable development (in the agriculture, forestry and fisheries sectors) conserves land, water, plant and animal genetic resources and it is environmentally non-degrading, technically appropriate, economically viable and socially acceptable.

NEED FOR SUSTAINABLE DEVELOPMENT

Aquaculture now accounts for roughly one third of the world's total supply of food fish and undoubtedly the contribution of aquaculture to sea food supplies will increase in the future. Aquaculture has potential to become a sustainable practice that can supplement capture fisheries and significantly contribute to feeding the world's growing population. Aquaculture is the fastest growing sector of the world food economy, increasing by more than 10 per cent per year and currently accounts for more than 30 per cent of all fish consumed.

Aquaculture, in common with all other food production practices, is facing challenges for sustainable development. Most aqua-farmers, like their terrestrial counterparts, are continuously pursuing ways and means of improving their production practices, to make them more efficient and cost-effective. Awareness of potential environmental problems has increased significantly. Efforts are under way to further improve human capacity, resource use and environmental management in aquaculture. COFI emphasized enhancement of inland fish production through integrated aquaculture-agriculture farming systems and integrated utilisation of small and medium-size water bodies.

Unsustainable aquaculture will only generate short and medium term profits for multinational corporations at the expense of long-term ecological balance and social stability. An unsustainable aquaculture development could exacerbate the problems and create new ones, damaging our important and already stressed coastal areas. Sustainable development alternatives are needed to ensure that in the future aquaculture can contribute to the growing need for seafood products. The sustainable development includes- "the management and conservation of natural resource base, and the orientation of technological and institutional change in such a manner to ensure the attainment and continued satisfaction for present and future generations. Such developments conserve land, water, plant and genetic resources as well they are environmentally non-degrading, technologically appropriate, economically viable and socially acceptable. The promotion of sustainable aquaculture development requires that "enabling environments", in particular those aimed at ensuring continuing human resource development and capacity building, are created and maintained. The FAO Code of Conduct for Responsible Fisheries contains principles and provisions in support of sustainable aquaculture development. The Code recognises the Special Requirements of Developing Countries, and its Article 5 addresses in particular these needs, especially in the areas of financial and technical assistance, technology transfer, training and scientific cooperation. There are a number of alternatives for sustainable development of aquaculture which include ecological aquaculture, organic aquaculture, composite fish culture, integrated aquaculture and closed recirculatory systems.

ECOLOGICAL AQUACULTURE

Ecological aquaculture has been defined as- "an alternative model of aquaculture research and development that brings the technical aspects of ecological principles and ecosystem thinking to aquaculture and concerns for the wider social, economic and environmental context of aquaculture".

There are few main principles of aquaculture:

- To preserve the form and function of natural resources
- To ensure trophic level efficiency
- To use native species so as not to contribute to biological pollution
- To share the practices and information on a global scale
- To ensure that system is integrated into the local economy and community in terms of food production and employment

Ecological aquaculture focuses on the development of farming systems that protect the environments in which they are situated and enhances the quality of these environments while at the same time maintaining a productive culture system.

ORGANIC AQUACULTURE

Sustainability is one of the main goals of organic food production.

Some of the basic principles of organic aquaculture according to the International Federation of Organic Agriculture Movements are as follows:

- To encourage natural biological cycles in the production of aquatic organisms
- Using various methods of disease control
- No use of synthetic fertilizer or other chemicals in production
- Use of polyculture technologies whenever possible

POLYCULTURE AND INTEGRATED AQUACULTURE

Polyculture and integrated aquaculture are methods of raising diverse organisms within the same farming systems, where each species utilises a distinct niche and distinct resources within the farming complex. This may involve the rearing of several aquatic organisms together or in conjunction with terrestrial plants or animals. Polyculture system can provide mutual benefits to the organisms reared by allowing for a balanced use of the available aquatic resources while integrated systems can increase the economic efficiency through improved conversion rates of input materials. The waste from one organism is used as input to another resulting in the optimal use of resources and less pollution overall. Although still experimental, other systems such as-integration of sea weed, fish and abalone culture and polyculture of shrimp and tilapia, have proved to be ecologically efficient methods for growing a variety of organisms and may increase profit at fish farms.

RECIRCULATING SYSTEM

Concerns for water conservation and reduced waste discharges have realised the use of closed recirculating aquaculture systems. This system is made up of three basic components: culture chamber, settling chamber and biological filter. Water enters the culture chamber, flows through the settling chamber and then moves through the biological filter to remove additional particulate matter. The water is then circulated back through the system culture chamber.

Recirculating systems conserve water and allow control of environmental factors (temperature, salinity and oxygen), predators and introduction and transfer of diseases. This system has less impact on environment because of their close nature - wastes and uneaten feed are not simply released in the ambient environment. In recirculating system, wastes are filtered out of the culture system and disposed of in a responsible manner.

In order for aquaculture to develop into an environmentally and socially responsible food production endeavour, following points should be recommended:

- Implement more ecologically sustainable practices
- Transition to use of closed systems and low discharge systems, especially those that provide total containment of fish and recovery or reuse of wastes
- Significantly decrease or eliminate the dependence on wild fisheries
- Develop sustainable aquaculture operations that provide long term social and economic benefits to communities

ISSUES OF SUSTAINABLE DEVELOPMENT

Sustainable development and sustainability are complex issues that are difficult to define and apply to aquaculture. The "systems approach", however, can assist understanding of these issues, as they relate to aquaculture development.

The term sustainability has been defined in various ways but perhaps the most widely used is based on the definition of "sustainable development" in the Brundtland report: "sustainable development is development that meets the needs of the present without compromising the ability of future generations to meet their own needs". An even more succinct definition is that of the International Union for the Conservation of Nature (IUCN): "sustainable development improves people's quality of life within the context of the Earth's carrying capacity". These definitions contain two key concepts: meeting the present and future needs of the world's people, and accepting the limitations of the environment to provide resources and to receive wastes for the present and for the future. The Food and Agriculture Organization of the United Nations (FAO), in particular, recognized that increased capacity at the national level is required to achieve sustainable development by including the need for "institutional change" in definitions of sustainable agricultural development. The recognition that institutions are important highlights the need for education and training, effective institutional arrangements and a legal and policy framework to underpin sustainable development of agriculture, and indeed aquaculture.

Sustainability is commonly split into three separate components: social sustainability (SS); economic sustainability (EcS); environmental sustainability (ES). Whilst social sustainability criteria are difficult to define, the definition of economic and environmental sustainability are providing a basis from which management options can evolve in aquaculture projects. To take sustainability to a more practical level requires consideration of environmental, social and economic issues in aquaculture development. Thus, the approach to sustainability implies a systems approach.

There are general guidelines available on the different issues to consider. The Code of Conduct on Responsible Fisheries (CCRF), adopted by the FAO Conference in 1995 in particular, identifies a number of key issues. The Code sets out principles and international standards of behaviour for responsible practices, to ensure effective conservation, management and development of living aquatic resources while, at the same time, recognizing the nutritional, economic, social, environmental and cultural importance of fisheries and aquaculture, and the interests of all those involved in these sectors. The Code's Article on Aquaculture Development contains provisions relating to aquaculture, including culture-based fisheries. Fundamentally, the Code recognizes the importance of activities that support the development of aquaculture at different levels:

- The producer level;
- The local area, *i.e.*, the farm and its integration into local area management and rural development schemes;
- The national institutional and policy environment; and
- International and transboundary issues.

The Code identifies many key principles in development of management strategies based on an understanding of aquaculture systems-from the farm to national and international levels. It also provides a basis for a systems approach.

THE SYSTEMS APPROACH

Farm Level

There is a lot of information on aquaculture farming systems, and various definitions are available, such as the level of intensity of management and output, and degree of integration with other on-farm activities. A considerable literature exists on integrated (agriculture-aquaculture and vice versa) farming systems (Edwards, 1998 and "Farming species and systems" in these proceedings).

However, there are a wide range of culture species, culture facilities and management practices in use, and thus a very wide range of farming systems.

Key factors to be understood in the functioning of a farming system are the technologies of production and social, economic and environmental aspects. At the technology level, feeds, feed additives and fertilizers, water quality, seed quality and availability, chemoterapeutants and other chemicals, disposal of wastes that may adversely affect human health and/or the environment, and food safety of aquaculture products all require consideration. It has also been emphasised that a better understanding of the microbial populations in aquaculture systems, their interaction with the health of the farmed animals, and their role in maintaining a healthy aquatic environment are required.

The systems approach at the farm level can be used to understand and improve the efficiency of use of key natural resources-particularly water, nutrients, land, seed and financial resources. When focussing on the mix of all sustainability criteria through a systems approach, we have to deal with such factors as: appropriate densities, production and husbandry systems geared to the animal's health, maintaining ecological balance within the pond or other growout habitat, and provision of optimal social and economic benefits. It is inevitable that impacts on natural resources will become an increasingly important issue in the new millennium, and the systems approach can be used to analyse and develop the solutions required for more sustainable use of natural resources.

A wide range of management systems is already employed in aquaculture operations with varying degrees of success. Given that aquaculture systems range from small, relatively self-contained farms for subsistence, to large-scale commercial units for trade purposes, variable success is hardly surprising. Thus, a "one-size fits all" approach is unlikely to be successful. A systematic approach to production management, however, allows the farmer to manipulate and control production inputs that will result in more efficient, cost-effective production and minimize excessive outputs with negative environmental impacts. There is a tremendous body of information on site selection, farm construction and design features, aquatic animal health management, broodstock and seed production and care, production techniques, the use of appropriate feeds, feed additives and fertilizers, water and sediment management, including effluent control, and other topics.

The challenge is to optimize dissemination and use of such information and experience.

The systems approach can also be used for aquatic animal health management. Here the emphasis needs to move more towards management procedures, policies and products that can prevent or effectively eradicate significant pathogens, prevent re-infection through contaminants, and manage diseases in an environmentally sustainable manner. Subasinghe *et al.* (1998) provide a discussion of the role of the systems approach in aquatic animal health management.

The Local Level

The systems approach at the local level recognizes that individual farms cannot be seen in isolation, and that there are many interactions between an aquaculture farm and the external environment-including environmental resources and local communities. Furthermore, there can be significant cumulative effects where there are large numbers of farms crowded in small areas.

Environmental interactions with aquaculture arise from a wide range of inter related factors including availability, amount and quality of resources; type of species cultured; size of farm; culture systems management; and environmental characteristics of the farm location. Environmental interactions are not limited to impacts of aquaculture on the environment, but include environmental impacts on aquaculture and impacts of aquaculture on aquaculture. Perhaps less well known or documented are the many ways that aquaculture can contribute to environmental improvement, for example mollusc farms' desedimentation or improved nutrient turnover, or water storage on small-scale freshwater farms.

At the local level, social and institutional interactions are also important and need to be better understood, for example:

- Participation of, and benefits to, rural communities;
- Institutional support through extension services;
- Access to information etc.

A systems approach attempts to understand these linkages and develop management strategies based on such understanding.

The future of integration of aquaculture into local ecological and social systems requires more focus on local area development planning. Fortunately, increasing attention is being given to such issues, particularly in coastal areas. Integrated coastal management (ICM) is a process that addresses the use, sustainable development and protection of coastal areas, and according to GESAMP (1996) "comprehensive area-specific marine management and planning is essential for maintaining the long-term ecological integrity and productivity and economic benefit of coastal regions". ICM is made operational through such activities as:

- Land use zoning and buffer zones;
- Regulations, including permitting to undertake different activities;
- Nonregulatory mechanisms;
- Construction of infrastructure;
- Conflict resolution procedures;
- Voluntary monitoring; and
- Impact assessment techniques.

More participatory approaches to planning of aquaculture development will also be given attention with the move towards integrated development planning. Practical experience in implementation ICM for aquaculture is limited, which is in large measure because of the absence of adequate policies and legislation and institutional problems, such as a lack of unitary authorities with sufficiently broad powers and responsibilities, as well as limited training and education of people concerned. In inland rural areas, increasing attention is being given to integration of aquaculture into rural development and special area management plans. Increasing emphasis is also being given to promotion of aquaculture for poverty alleviation. Such an approach requires emphasis on immediate social needs and people's livelihoods (and how aquaculture might meet these needs and contribute to improved livelihoods) rather than a technology/aquaculture driven approach.

The emphasis on aquaculture for development, rather than development of aquaculture may lead to some fundamental changes in the approach to promotion of aquaculture in the coming years.

National, Regional and International Levels

At the national level, government policy, and institutional and human capacity are most important in providing a strong foundation for aquaculture to develop in a su.

These issues are covered extensively in the Conference's policy session and are not discussed in detail here, except by recognition that most community and farm activities are influenced by national-level policy, legislation and institutional support. For example, the level of aquatic animal disease affecting small-scale producers or enhanced fisheries is related to national policies for quarantine and movement of live aquatic animals, which affect the risk of exposure of small-scale producers to serious aquatic pathogens. Inter-and intra-country trading patterns and movement of aquatic animals also affect these risks. International conventions (*e.g.*, the Convention on Biodiversity), trade and consumer preferences, all clearly impact aquaculture development at a local level.

CONTROLLING THE ENVIRONMENT OF LIVE MOLLUSCS

Growing and harvesting of molluscs should be restricted to areas free from direct sources of sewage pollution. This requires knowledge of local geography, prevailing water currents and the local treatment and discharge of sewage. Also monitoring of the microbiological quality of water is required. Thus the current standard for water quality in growing areas in U.S. is 14 MPN fecal coliforms/ 100 ml water, with no more than 10 per cent of samples exceeding 43 MPN fecal coliforms/100 ml. However, the value of fecal coliforms as indicator of contamination and possible presence of disease agents have serious limitations as already discussed.

Also the correlation between presence of indicator bacteria and various disease agents in water and in shellfish has been questioned. The concentration of microorganisms in filter feeding shellfish varies enormously from animal to animal and also depends on weather conditions, temperatures and general activity of the shellfish. For these reasons, there are no microbiological standards for water quality in growing areas specified by the European Economic Community (EEC). Instead, the EEC has placed a microbiological standard (EEC Directive 91/492/EEC) on shellfish for direct consumption :

<300 fecal coliforms/100 g meat or

<230*E.coli*/100 g meat (based on MPN-test)

absence of *Salmonella* in 25 g meat

PSP < 80ìg/100 g edible meat

DSP not to be detected by customary biological testing method.

It should be noted here that the regulatory agencies in US and Europe have returned to traditional control options (sampling, testing and comparing results with microbiological standards) in an attempt to provide safety in consumption of raw molluscs. These the methods do not provide any guarantees for safety, and by the application of HACCP-concept it is also clear that noguarantee of safety can be provided. This point should be made clear to consumers, who insist on consumption of raw molluscs. Such warnings are actually used in seafood restaurants in Florida, USA.

Also control of the environment for the presence of toxic dinoflagellates is difficult and facing some of the same types of problems as discussed for bacteria and viruses. The EEC requires periodical (weekly) sampling of water and shellfish from growing and harvesting areas, and in case of elevated presence of toxic algae, the fishing area will be closed. However, analytical techniques is one of the great problems.

An alternate means of securing safety of shellfish is by relaying or depuration and in a number of countries this is required by regulation. Depuration involves placing the shellfish in tanks with clean circulating seawater. Various methods can be used to disinfect the water such as ultra-violet light, chlorine, iodophors, ozone and activated oxygen and the shellfish is simply removed from suspect areas to waters which are known to be unpolluted. Both methods are of limited efficiency in the removal of viruses and vibrios from the shellfish. Efficiency has normally been verified by testing the animals for presence of *E. coli*. This organism is not suitable as indicator, however, and alternative method is required. Non-microbial contamination (toxins/biotoxins, heavy metals, petroleum, hydrocarbons, radionucleides, pesticides) depurate so slowly that commercial depuration is uneconomical.

In most countries control and monitoring of the environment is the responsibility of the governments and these should be consulted for detailed information. It is expected that governments will ban fishing or harvesting of molluscs, if criteria are not met.

Temperature Control

Time-temperature (Txt) conditions at all times from catching to distribution is a CCP-1 in preventing growth of pathogens and spoilage bacteria. Thus the time-lapse between each step in the flow diagram must be monitored and similarly the temperature of the environment, chill rooms, factory etc. as well as the temperature of the product must be recorded.

Factory Hygiene and Sanitation

Factory hygiene as well as personal hygiene and sanitation are CCPs in the prevention of contamination of products with microorganisms, filth and any other foreign material during processing. The seriousness (risk) of this hazard varies depending on local conditions (factory lay-out and design, facilities) and intended use of product (cooking or no cooking before being eaten). For this reason, a detailed description of the requirements must be produced in each individual case. These instructions must specify precisely when to clean and sanitize, how to do it, who is responsible, equipment and chemical agents to be used etc. This CCP can then be controlled and monitored by visual inspection of procedures and recording of data in check-lists. Occasionally a microbiological check of cleanliness of surfaces coming into direct contact with exposed meat can be made. Bacteriological control must be regarded more as a verification procedure than monitoring of a CCP. The frequency to carry out this type of check also depends on circumstances. In cases where changes in procedures or personnel has taken place. This control procedure must be carried out on a weekly or maybe on a daily basis. In other cases where routines are well established, microbiological control of cleanliness can be carried out monthly or maybe abolished altogether. Water quality is a CCP-1 in preventing contamination from this source. Monitoring of this CCP-1 can be carried out by microbiological testing. Where in-plant chlorination is used, chlorine levels must be measured and recorded. Chlorine levels should be measured daily with recommended levels of 2–5 ppm.

Fish Raw Materials for Further Processing

The hazard analysis of these products is fairly straightforward and uncomplicated. The live animals are caught in the sea or freshwater, handled and - in most cases - processed without any use of additives or chemical preservatives and finally distributed with chilling or freezing as the only means of preservation. Most fish and crustaceans are cooked before eating although a few countries such as Japan have a tradition for eating raw fish. The epidemiological records show that these products have caused a number of food poisoning outbreaks, but nearly all have been related to the presence of heat stable toxins (biotoxins, histamine).

Live fish and crustaceans and raw products may be contaminated with a number of pathogenic bacteria normally found in the aquatic environment such as *C. botulinum, V. parahaemolyticus*, various *Vibrio* sp., *L. monocytogenes, Aeromonas* sp. However, only the growth of these organisms can be regarded as a hazard, as pathogenicity is related to preformed toxin in the food (*C. botulinum*) or the minimal infection dose is known to be high (*Vibrio*). The severity of diseases related to these organisms may be high (botulism, cholera) or low (*Aeromonas infections*), but the likelihood of provoking diseases (risk) is extremely low. The pathogenic strains require temperatures > 1 ° C for growth and they are competing with the normal spoilage flora whose growth potential is comparatively much higher at low temperatures. Thus the products are likely to be spoiled before production of toxin or development of high numbers of pathogens. When the products are cooked before consumption, this will completely eliminate the risk. Pathogenic bacteria from the animal/human reservoir (*Salmonella, E. coli, Shigella, Staphylococcus aureus*) may contaminate the live animal depending on the fishing area and further contamination may take place during landing and processing. The diseases which these organisms can provoke are serious, but if numbers on the products are low (*i.e.*, no growth has taken place) the likelihood of this to happen (risk) is very low indeed. Cooking before consumption will eliminate the risk. However, an indirect hazard exists if contaminated products are polluting the working areas (industry, kitchen) and thereby transporting the pathogens to products which are *not* cooked before eating (cross contamination). This indirect hazard must also be prevented.

In contrast, the effect from growth of histamine producing bacteria (*Morganella morganii*) will not be eliminated by cooking or any other heat treatment as the heat resistance of histamine is high. The risk of histamine poisoning, if fish (*Scombroidae*) have been kept for some time at elevated temperatures (> 5 ° C) is therefore high.

Fish caught in certain areas may be infected with parasites dangerous to human health. The severity of the possible disease depends on the parasite involved, and the likelihood of contracting parasites from fish is eliminated if the fish is cooked before consumption. A low risk will exist if fish are consumed raw.

The presence of biotoxins and chemicals in fish depends on fish species, fishing area and season. The biotoxins are heat-stable and the risk of intoxication after consumption (raw or cooked) is high. The safety hazards related to fish raw materials for further processing and to consumption of fresh and frozen fish. Heavy contamination and in particular, growth of specific spoilage bacteria is certain to reduce the normal and expected shelf-life of the product (high risk). This may of course cause serious commercial problems but no lives are endangered. Thus, severity is low.

ENVIRONMENTAL EMERGENCIES AFFECTING FISHERIES

There are three types of environmental uncertainties affecting fisheries: (1) natural environmental oscillations, (2) global environmental change and (3) environmental emergencies.

Natural oscillations affect fish abundance and availability: seasonal oscillations are usually forecasted with reasonable accuracy and exploited by fishing industries whereas inter-annual oscillations that occur with a range of frequencies (such as 11 to 50-60 year cycles) are less predictable and are only slowly being understood (*e.g.*, El Niño and La Niña phenomena). Environmental emergencies correspond to environmental crises emerging as a consequence of an unforeseen combination of circumstances with results that call for immediate action in the form of contingency plans or mitigating action, such as assistance or relief. These include fish kills, harmful algal blooms and oil spills (caused by tanker accidents or from incidents related to offshore development).

FISH KILLS

Massive fish kills affecting large biomasses across relatively large aquatic areas are often triggered by unpredictable environmental conditions resulting from accidental (or sometimes illegal) events such as the release of toxic substances from industrial waste storage facilities (*e.g.*, following a fire or mine tailing impoundment after heavy rains or from malfunctioning waste treatment facilities).

As well, heavy winds, resulting in up-welling of oxygen-depleted bottom water, can be the cause of fish kills in stratified coastal waters and great lakes. The fish kills themselves may aggregate oxygen depletion as large amounts of dead organisms accumulate on the bottom. Fish kills can also result from cold spikes (*e.g.*, in the Northern Adriatic) when exceptionally cold atmospheric conditions reduce water temperature in coastal areas, provoking massive kills of sessile or benthic populations (*e.g.*, Norwegian lobsters).

Harmful Algal Blooms

About 300 hundred species of micro-algae can form mass occurrences at times, nearly one fourth of which produce toxins. Scientifically referred to as

'Harmful Algal Bloom' (HAB), they are commonly called "red tides" as the large patches of planktonic concentration may appear red on the surface. Such proliferation of micro-algae - thought to be activated by environmental changes - in marine or brackish waters can cause massive fish kills, contaminate seafood with toxins and alter ecosystems in harmful ways.

A broad classification of HABs distinguishes two groups of organisms: the toxin producers, which can contaminate seafood or kill fish, and the high-biomass producers, which can cause anoxia and indiscriminate kills of marine life after reaching dense concentrations.

The impact of harmful micro-algae is particularly evident when marine food resources, such as mussels grown in aquaculture facilities, are affected. Shellfish, and in some cases finfish, are often not visibly affected by the algae but non-etheless accumulate the toxins in their organs. The toxins may subsequently be transmitted to humans through consumption of contaminated seafood and become a serious health threat.

Tanker Accidents

Tankers carry about 1 000 million tonnes of oil per year, about 50 per cent of which is extracted on the continental shelf. On some offshore oil fields shuttle tankers are the main way of delivering hydrocarbons to the onshore terminals. The principal causes of tanker accidents that lead to large spills include running aground and into shore reefs, collisions with other vessels, and cargo fires and explosions. More recently, a number of accidents occurred when vessel structures collapsed during severe weather conditions.

Tanker accidents often result in vast oil spills or massive release of harmful chemicals transported in bulk. According to the International Maritime Organisation (IMO), the amounts of oil spilled during tanker accidents in 1989 and in 1990 were 114 000 and 45 000 tonnes, respectively. Some of the most substantial oil spills have been: the Torrey Canyon in 1967 with 95 000 tonnes of oil spilled on the French and British shores; the Amoco Cadiz in 1978 with 220 000 tonnes of oil spilled, the Exxon Valdez in 1989 with 40 000 tonnes and the Braer in 1993 with 85 000 tonnes of oil spilled.

The most dangerous are accidents involving underwater storage tanks that contain dangerous substances such as methane. Such incidents will be possible in the area of Shtokmanovskoe field developments in the Barents Sea where over 3 000 tonnes of methane are planned to be stored underwater. The resulting spills could have disastrous effects on coastal ecosystems, fisheries, aquaculture installations and, in some instances, on human life. Major impacts of oil spills on fisheries and aquaculture are the smearing of nets and fish cages and the tainting of fish and shellfish, rendering them unfit for marketing. Longer-term impacts on the ecosystem depend on the nature of the pollutant and the ecological characteristics of the area.

Pipelines Accidents

Complex and extensive systems of underwater pipelines, amounting to thousands of kilometres, are among the greatest environmental risk factors during offshore oil developments, along with tankers and drilling operations. Causes of accidents range from defective material, pipe corrosion, ground erosion, tectonic movements, ship anchors and bottom trawls. Depending on the damage, the pipeline may become a source of leakage or of an abrupt (even explosive) blowout of hydrocarbons near the bottom. The intensity and scale of toxic impacts on the marine biota in the accident zone varies depending on a combination of many factors. It is important to consider that, in a number of cases, the accidental oil and gas spills and blowouts of mainland pipelines threatens coastal marine ecosystems. This can occur when such accidents take place, for example, near big rivers where river pollution eventually affects the sea zone near the river mouth.

Drilling Accidents

Drilling accidents are usually associated with unexpected blowouts of liquid and gaseous hydrocarbons from the well. Only tanker oil spills compete with drilling accidents in regularity and severity. Some rare occurrences could lead to catastrophic situations involving intense and prolonged hydrocarbon gushing. Often they involve hydrocarbon spills and blowouts during drilling operations. Usually, these accidents do not attract any special attention. Non-etheless, the ecological hazards and associated environmental risks can be considerable - especially given their frequency - and could ultimately lead to chronic impacts on the marine environment.

Prevention of Emergencies

Precaution In fisheries and aquaculture, precaution is an indispensable and vital attitude. Many activities involved with fishing and farming fish involve a very high degree of risk. Anticipating threats and trying to avoid damage to equipment and installations, to protect farmed stock, or to prevent harm to owner or the consumers' health is in the interest of fishers and fish farmers. While fishers and fish farmers are used to dealing with highly variable environmental and climatic conditions, it is also known however that in many cases of climatic or environmental emergencies, which have affected fisheries, significant damage and economic loss could have been prevented by forethought, preparation and vigilance.

Solution

Being informed and prepared reduces the impact of surprise. Precautionary action, to be effective in emergencies, requires assessment of risks, including identification of uncertainties, definition of priorities (including priority resources, priorities for protection), and clear response strategies, *i.e.*, setting priorities of action, and their timing and location, which may involve aspects of organization and management, communication, equipment and manpower, training and exercises, public awareness campaigns, etc. In fisheries, generally, there is scope for developing and improving emergency preparedness and contingency planning.Not only fisheries, but many sectors deal with emergencies, risks and uncertainties, and approaches of risk management and contingency planning have been developed, for example, in the realms of natural disaster planning and mitigation, food safety and public health assurance, environmental hazard assessment and management, insurance business, etc. Fisheries stakeholders can learn from experiences made by others.

Action

Fisheries insurance In general, major risks confronting fishers and fish farmers can be divided into following general categories: asset risks, production and management risks, and market risks and personal and health risks.

Asset risks include loss of or damage to fishing vessels, equipment, and gear and aquaculture installations, as a result of natural or man-made disasters.

Production and management risks involve the loss of catch, production failure and fish disease. Market risks relate to changes in the prices of outputs and inputs, as well as increases in interest rates. Personal and health risks include accidents at sea and death and job-related illnesses. Of significance can be changes in consumer behaviour, especially if fish and fishery products are being associated with potential harm to human health, social well-being and environmental integrity.

There is increasing experience with developing and implementing fisheries insurance schemes. However, overall, there is still significant scope for further ypromoting such schemes, particularly in developing countries.In view of the high level of risk and the many types of risks associated with fisheries and aquaculture, private insurance companies are still reluctant to get involved in this sector, particularly, in traditional small-scale fisheries and fish-farming activities, and especially those in developing countries. However, fisheries insurance should be viewed from the perspective of the small-scale fishers and fish-farmers, which comprise the majority of the fishers population in developing countries. Such fishers need protection against losses caused by natural and man-made disasters. Fisheries insurance schemes are therefore indispensable and government support is necessary for their establishment.

The advantages and benefits provided by fisheries insurance vary from scheme to scheme. However, in general, the principal benefits of fisheries insurance are:

- Protecting fishers and fish-farmers against accidents and natural hazards beyond their control;

- Providing basic compensation for the loss of or damage to fishing vessels, gear and catch (or harvest), thus contributing to stabilisation of incomes within the fisheries sector;
- Reducing the risk to financial institutions, which provide credit to fishers and fish-farmers, in relation to fisheries credit;
- Reducing the risk for fishers and fish-farmers in investing their own resources in the adoption of new technologies and acquiring improved equipment;
- Fostering mutual assistance and co-operation among fishers, fish-farmers and their organizations;
- Reducing the unpredictable burden on government of providing emergency assistance in the wake of natural disasters;
- promoting stability in fishery enterprises and contributing to the general welfare of fisheries communities;
- Stabilizing the contribution of the fisheries sector to national economy.

Difficulties and problems that arise with respect to the introduction of fisheries insurance schemes include the following:

- Limited financial resources;
- Inadequate benchmark data on the extent of damage to vessels, gear, catches and production, ponds and installations;
- A wide variety of diverse fishing and aquaculture practices;
- Little apparent demand for insurance;exclusion of small-scale fishers and fish-farmers from insurance;
- Lack of well-established village institutions, such as co-operatives, to act as insurance agents;lack of legal framework for fisheries insurance;
- Lack of related government policy;
- Difficulty in promoting insurance policies, designing sustainable insurance programmes and co-ordinating the work of the agencies concerned;
- Difficulty of covering credit projects under insurance schemes;
- High cost of premiums and providing appropriate levels of benefits;
- Speed and transparency of claims settlement;
- Lack of trained personnel;
- Lack of understanding of the value of insurance; and
- Lack of reliable actuarial data for establishing premium rates.

Examples of contingency planning and emergency response measures

Measures to mitigate the impact of hurricanes and cyclones on fisheries may include:

- Assessment of risk and vulnerability to, and the economic implications of, storm damage to fisheries, highlighting weaknesses and strengths in existing practices and successful attempts to protect boats, fishing gear and infrastructure;

- Development and dissemination of simple methods of improving land protection, minimizing erosion, introduction of more storm-resistant agricultural crops and planting of forestry windbreaks or shelterbelts;
- Development and dissemination of improved measures and techniques such as infrastructure construction (*e.g.*, storm shelters, fishing jetties and mooring buoys);
- Installation of storm warning systems including radio relay stations, shore-to-boat and boat-to-boat communication networks;
- Supply of life-saving appliances; boat modification and motorization programmes;
- Establishment of local sea and storm safety committees and an effective search and rescue capability, and provision of training and technical advice on sea safety;
- Demonstration and training programmes, including national and sub-regional workshops on hurricane/cyclone impact mitigation, and production of training programmes and materials for farmers, forestry staff, fishers and extensionists.

In case of oil spills affecting or threatening fisheries and aquaculture operations, response options may include application of protection and clean up techniques, *e.g.*, using booms and other physical barriers, sorbent materials, and dispersants at safe distance from fish stock. Fishery closures can be imposed after an oil spill in order to prevent or minimise fishing gear contamination and to protect or reassure seafood consumers. Fishers can agree to a voluntary suspension of fishing activity as a precautionary measure during a period when oil is drifting in their normal fishing area, and thereby avoid repeatedly contaminating fishing gear.

Aquaculture-specific options may include moving floating facilities out of the path of slicks, sinking of specially designed cages to allow oil to pass, transfer of stock to areas unlikely to be affected, reduction or suspension of feeding, temporarily suspending the replenishment of seawater, in the case of land-based installations. In some cases aquaculture operators may face the risk of ultimately losing all the stock due to oil spill damage. Harvesting before the stock becomes oiled might be possible, albeit selling the products at a lower price, and thereby salvaging some of its value. Conversely, normal harvesting could be delayed to allow contaminated stock to depurate and become taint-free.

In an oil spill it is vital to communicate information to the media and the public in an effective manner on the likelihood of adverse consequences for fishery resources. Inaccurate public information about tainting and contamination may limit the range of management strategies available, causing unnecessary fishing and harvesting restrictions and/or loss of consumer confidence in the market. Risk communication is an ongoing process that must be addressed in both spill response planning as well as during the spill event.

Information about risk can be communicated through a variety of channels, from media reports to public meetings.In the case of harmful algal blooms affecting aquaculture operations countermeasures at farm level may ideally include:

(i) Provision of multiple water intakes (*e.g.*, pumping of plankton-free water from various depths);

(ii) Vertical movement of cultured organisms (*e.g.*, sinking of cages);

(iii) Relocating of culture units to unaffected areas which requires detachable moorings and towable structures;

(iv) Pre-emptive harvesting;

(v) Reduction of food supply and stress to lower metabolism of cultured stock;

(vi) On-site shielding of stock (*e.g.*, bubble curtains, injection of clear water, non-porous barriers);

(vii) Cortisone treatment to reduce gill-swelling.

These are options that realistically would require significant investment, which in many cases is rather unlikely given the socio-economic living conditions of many fish farmers. Examples of contingency planning in aquatic animal health management are described in Emergencies in aquaculture - disease outbreaks.

Outlook

Governments may have a role to play in the promotion of aquaculture production and fishing activities in unpolluted waters and low risk areas that should be combined with hazard assessment and regular monitoring of environmental conditions. In general, systematic collection of information on apparent pollution events or other patterns of environmental deterioration may prove very useful for various purposes such as early detection/early warning schemes (*e.g.*, on oil spills, phytoplankton blooms); improved demarcation of pollution-exposed areas (*e.g.*, changing patterns of distribution/expansion of contamination); and record-keeping of chronology/history of events (frequency, duration, time of year, etc.).

The FAO Fisheries Department is committed, in its medium-term programme 2000-2005, to provide member countries with information and indicators to better forecast emergencies and disasters in fisheries and aquaculture. In particular, the activities will aim at the promotion of ecosystem/ environmental information and management by developing, testing, documenting and disseminating approaches, methodologies, models and procedures for effectively utilizing information on environmental variation and biological community composition in its applications to fishery management and industrial and small-scale fisheries. The work would also include advice on ecological and environmental applications to members' countries, and regional fishery bodies.

ENVIRONMENTALIST ADVOCACY AND FISHERIES MANAGEMENT

Environmentalists and fishermen rarely agree on anything. It's a pity, because conventional logic would say that, fundamentally, environmentalists and fishing people should stand shoulder-to-shoulder in a common cause for maintaining resources *sustainability*, while blending that with their sensible usage. This, especially, with respect to the damage to the environment due to pollution and coastal habitat destruction. Unfortunately, only too often over-eager environmentalist see fishing people as enemies of the environment, and a controversy between the two, at times on low fire, at times raging, is going on in many countries, and even globally.

For example, to some environmentalists *sustainability* means reducing commercial fishing to no more than a very small inshore one-man vessel industry with the interests of seaweed/fauna growing on the seabed being preferable to the development of an industry that over the generations has been producing top quality nutritional food. But, they say it's needed, for it is only by reducing the capacity of fishing fleets that a solution to the problem of over-exploitation of the marine environment can be found. "The fishing industry cannot exist if there are no fish left" – is the argument. They believe that reduction of excess fishing capacity, creation of "no-take zones" covering about 20 or more per cent of marine habitats, and political enforcement of sustainable fishing levels would solve the problem of stocks impoverishment. But what they fail to see or often ignore are the dozens of toxic chemicals and excess nutrients that abound in upstream water and air pollution and their joint effect on coastal marine areas, and that the most deadly pollution that affects coastal habitats including fish nursery and feeding grounds, comes from industrial sources and, in particular, from the petro-chemical and electro-chemical sectors.

Other factors ignored by the environmentalists' advocacy are the crucial influence of climatic and oceanic processes and fluctuations on fish populations, well documented in the history. Species with narrow temperature preference limits, are affected by thermal anomalies that delay or hasten spawning and hatching, and displace spawning and feeding grounds. Survival of larvae and juveniles depends, apart from hydrographic conditions, also on availability of the right food, at the right place and time, as well as on the rate of predation. Changes in river flows raise or reduce salinity in estuaries, deplete and displace fish and aquatic plant species. Climate variability is the key controlling factor in fishing yields for about half of the world's large marine ecosystems, including the East and West Greenland shelves, the Benguela Current off Southwest Africa, the Canary Current off Northwest Africa and the Humboldt Current off the west coast of South America. Over the past decades, modes of oceanic variability such as the Pacific Decadal Oscillation and the North Atlantic Oscillation have been related to shifts in marine ecosystem structure, species composition, distribution and biogeochemical processes. Thus, since patterns of species abundances shift with the ocean climate, when climatic, biological and oceanographic conditions are just right, fish can respond with an extremely strong year class, or a series of them, and the other way around.

FISHERIES MANAGERS AND ECONOMICS

Environmental lobbying encourages fisheries managers to think that if they got hold of controlling the fishing, all their problems would be solved. Ignoring anthropogenic and natural environmental factors, which critically affect natural mortality is ridiculous. But to many of them all these ecological problems are only a noise, which creates confusion in their favored system. In that they're joined by some economists, who analyze fisheries in purely economic-financial terms, with hardly paying even lip service to non-fishing factors that affect the whole system, fishery resources included. A neo-liberal economic approach inserted the agenda of privatization of fishery resources as a medicine to overfishing. This agenda is often disguised as rights-based management. Politically it depends on who holds or is offered the rights, and economically-are those rights marketable, a solution promoted by neo-liberal economists.

Accordingly, the quota system has been introduced to fisheries management, particularly the individual transferable quotas (ITQ), which had great theoretical appeal to those economists. They embraced ITQ-system quickly and uncritically, and appeared to regard it as a cure-all for fisheries management problems. While the overall benefits of the ITQ-system are highly disputable, what is not is that it causes an ongoing concentration of quota in the hands of a smaller number of larger operators and a resulting displacement of fishermen owner-operators. Thus, the introduction of the ITQs has become also a political tool for redistribution of benefits from marine fishery resources away from the small to the hands of large firms and corporate interests. Notwithstanding, the whole idea of rationalization of fisheries through privatization is based on misperception and, hence, it fails to achieve the desired results of sustainability, all the more that for most fisheries any promise of true privatization is a hoax. Evidence accumulated over many years shows that there's more than a single viable resource management option. Private, state, and communal property and fishing rights are all potential options and there's no panacea for fisheries management. Fishery laws and regulations should be separately tailored for every fishery and environment, and for the different socio-economic conditions. The tunnel vision of some economists, which disregards external costs and benefits, represents achievement of the highest profits and free marketing of fishing rights and quotas as also the highest social and national benefits. Often, consequences of such approach becoming actual policy have been social and economic devastation of fishing communities and dislocation of fishing people.

8

Cooperation of Aquaculture

AQUACULTURE

Aquaculture -- also known as fish or shellfish farming -- refers to the breeding, rearing, and harvesting of plants and animals in all types of water environments including ponds, rivers, lakes, and the ocean. Researchers and aquaculture producers are "farming" all kinds of freshwater and marine species of fish, shellfish, and plants. Aquaculture produces food fish, sport fish, bait fish, ornamental fish, crustaceans, mollusks, algae, sea vegetables, and fish eggs. Aquaculture includes the production of seafood from hatchery fish and shellfish which are grown to market size in ponds, tanks, cages, or raceways. Stock restoration or "enhancement" is a form of aquaculture in which hatchery fish and shellfish are released into the wild to rebuild wild populations or coastal habitats such as oyster reefs. Aquaculture also includes the production of ornamental fish for the aquarium trade, and growing plant species used in a range of food, pharmaceutical, nutritional, and biotechnology products.

An understanding of the principles of operation of capture and culture fisheries helps to throw light on the definition of aquaculture. The expressions capture and culture fisheries are self-explanatory. In the former, one reaps the aquatic harvest without having to sow, whereas, in the latter, one has to sow the seed, nurse it, tend it, rear it and harvest it when it grows to marketable size.

Examples of capture fisheries are the natural fisheries of the seas, estuaries, rivers, lagoons, large lakes etc.

Culture fisheries are usually carried out in small water bodies which can be manipulated, pre-prepared for stocking; which are often manured and/or fertilized before, during and after stcking; and/or where fish are fed from extraneous sources. Pen culture, cage culture, culture in running waters, in recirculating systems and in reconditioned water are special types of aquaculture. All shades of intermediate stages between true capture and culture fisheries exist such as in man-made-lakes, which are stocked extraneously but where no manuring, fertilizing and feeding are generally done. Stocking is often done in large water-bodies such as lagoons and rivers where natural stocks have undergone 'depletion'.

The principles of management of capture and culture fisheries are very different from each other. In the case of capture fisheries one has to attempt to harvest maximum sustainable yield by regulating fishing effort and mesh after taking into account parameters of population dynamics such as rates of recruitment, natural and fishing mortalities, fish growth and size at which recruitment occurs. Management of capture fisheries requires knowledge of the dynamics of the fish populations under exploitation. The extended exclusive economic zone of 200 miles brings into focus the national and international complexities of regulating the capture fisheries of the seas and the oceans and apportionment of the marine harvest because fish populations do not abide by man-made boundaries. In the case of culture fisheries, no detailed knowledge of the population dynamics of the cultivated finfish or shell fish is involved. Here, one has to breed, if one technically can, the chosen fish under controlled conditions, if it does not breed naturally, and develop fish husbandry practices so as to be able to formulate economically viable technologies. For effective aquaculture, one has to gain familiarity and control water quality to enhance its biological productivity; one has to understand fish nutrition so as to be able to formulate nutritionally balanced fish diet; one has to delve deep into fish genetics so as to be able to evolve new varieties and strains which bestow commercial advantages to the product in terms of superior growth rate, nutritive value, bonelesness, taste, odour etc.; one has to prevent incidence of fish infections and diseases through prophylatics and therapeutics.

With this background information, a definition of aquaculture can be attempted. Aquaculture has been defined by the Japanese Resource Council, Science and Technology Agency as under: "Aquaculture is an industrial process of raising aquatic organisms upto final commercial production within properly partitioned aquatic areas, controlling the environmental factors and administering the life history of the organism positively and it has to be considered as an independent industry from the fisheries hitherto."

Aquaculture is organised production of a crop in the aquatic medium. The crop may be that of an animal or a plant. Naturally, the organism cultured has to be ordained by nature as aquatic.

Examples are:

Finfish: Tilapia, carp, trout, milkfish, bait minnow, yellow tail, mullet, cat fish.

Shellfish: Shrimps, prawns, oysters, mussels, pearl oyster for cultured pearls (eg. Japanese pearl oyster, Pinctada fucata).

Plants: Water chestnut (Trapa natans). Red alga of Japan, "Norie" (Porphyra). Red alga of Philippines & U.S.A. (Eucheuma) Brown alga of Japan, "Wakame" (Undaria).

During the last decade or so there has been noticeable a global upsurge for aquaculture. Some of the factors which have contributed to the upsurge are:

Increased and continuously rising cost of fishing operations due to steep rise of the price of fuel. Fear of reduction in marine fish landings by countries that depend on fishing in the territorical waters of other countries as a result of the new laws of the sea of 200 miles exclusive economic zone.

Need, in some countries, for finding alternative and/or additional employment for large numbers of surplus fishermen or under-employed farmers.

A persistent demand in most developed countries for high cost species like shrimps and prawns. This has greatly promoted interest in aquaculture in countries that wish to increase their foreign exchange earnings.

The behaviour of one of the world's most productive capture fisheries viz. that of Peruvian anchovy has, I believe, given a hard blow to aquaculture on the one hand, in making scarce and increasing the cost of fishmeal, which is the ingredient of most fish feeds, and, on the other, seeing the helpnessness of man in countrolling natural causes of fluctuations of marine fish yields, has created a desire in him to acquire control on processes of production through aquaculture.

Factors which have been unfavourable to the development of aquaculture are:

Shortage of fertilizers in most developing countries and their allocation to agriculture. In this respect, there is a measure of conflict between agriculture and aquaculture. Increasing prices and even the availability of fish meal, which, as stated earlier, is the ingredient of most fish-feeds. This is linked with the Peruvian Anchovy crisis, which, apart from aquaculture, adversely hit agriculture, through scarcity of guano and fertilizer, and poultry industry through scarcity of fish meal. This has led to search for cheaper protein substitutes in fish feeds and spurt of research activity in that direction in different countries.

While a general global environmental consciousness has ameliorated aquatic pollution and has thus helped fish culture, aquaculture itself is considered by some as a polluting agent, through release of water containing fish metabolites leading to eutrophication in the recipient waters, which may be a stream or a river or another kind of natural water-body. Discharge regulations which are applicable to aquaculture by authorities in some countries. The basic fact is that fishes in general help to keep the aquatic environment clean through exercising biological control of vectors (eg. of water-borne diseases like malaria, filaria etc). Aquaculture water and pond bottom mud often act as fertilizers to agricultural fields. Rarely does aquaculture discharge-water cause pollution.

Authentic proof it required to establish that aquaculture is a polluter. In whichever case, if it is proved that aquaculture has polluted the environment, the discharge water from aquaculture establishment would need to be treated and rendered innocuous before release.

Aquatic pollution, through discharge of agricultural pesticides, domestic wastes, trade effluents and oil spills, has very adversely affected aquaculture. In this respect, there is a measure of conflict between agriculture, especially cultivation of high yielding varieties (HYV) of cereals, and aquaculture. *E.g.*, cases of fish kills in streams and other water-bodies where pesticides fall or where industrial effluents are discharged and adverse effect on oyster beds off Japanese, U.S.A. and French Coasts. The well known cases of oil spills are those of the tankers: Tory Canyon (1967) and Amoco Cadiz (1978).

Absence of a constitutional provision for aquaculture as a discreet national activity and legal frame-work for governing its development and administration in most of the countries of the world are standing in the way of entrepreneurs making investment in aquaculture. Multi-disciplinary and systems characteristics of modern aquaculture need to be especially emphasised in a lecture on definition of aquaculture. Mention has been made earlier of some of the essential components of aquaculture such as water quality control, fish breading, fish genetics, fish nutrition, fish feed formulation, fish pathology, fish parasites and predator control etc. An aquaculturiest has to successfully carry out a whole series of operations before be is able to market his produce.

Complete package of practices have to be developed which involve accomplishment of several steps such as fish multiplication, nursing, tending, and rearing the young, all of which require special food for the larvae and the young fish; then growing the young to marketable size which require special feed again and often intensive feeding for quick growth.

The quality of fish feed would naturally depend on the species cultivated. All the above mentioned steps in the practice of aquaculture require rigid water quality control. The cultivated fish has to be saved from the depredations of predators all along its culture. The health of the fish has to be continuously monitored and guarded against infections and infestations which have got to be checked. The systems approach stands in contrast with disciplinal studies where a scientist take s.up a specific problem and goes deep into it to investigate a certain phe no menon or seeks to establish cause and effect relationship. Even the latter, depending on the nature of the problem, may be multi-disciplinary but it need not always necessarily be. A biochemist, for example, can effectively study fish nutrition and feed components of fish required at different stages of its life but, for successful aquaculture, the whole system involving scores of aspects, some of which have been high-lighted above, have to be worked out.

OBJECTIVES OF AQUACULTURE

Having defined aquaculture and mentioned some of the reasons which have contributed to imparting a fillip to aquaculture in recent times, it is proper to state the objectives of aquaculture. These are:

- Production of protein rich, nutritive, palatable and easily digestible human food benefiting the whole society through plentiful food supplies at low or reasonable cost.
- Providing new species and strengthening stocks of existing fish in natural and man-made water-bodies through artificial recruitment and transplantation.
- Production of sportfish and support to recreational fishing.
- Production of bait-fish for commercial and sport fishery.
- Production of ornamental fish for aesthetic appeal.
- Recycling of organic waste of human and livestock origin.

Land and aquatic resource utilization: this constitutes the macro-economic point of view benefiting the whole society. It involves (a) maximum resource allocation to aquaculture and its optimal utilization; (b) increasing standard of living by maximising profitability; and (c) creation of production surplus for export (earning foreign exchange especially important to most developing countries).

Providing means of sustenance and earning livelihood and monetary profit through commercial and industrial aquaculture. This constitutes the micro-economic point of view benefiting the producer. In the case of small-scale producer, the objective is to maximise income by greatest possible difference between income and production cost and, in the case of large scale producer, by maximising return on investment.

Production of industrial fish. Fish flesh, on the average, contains: moisture and oil, 80 per cent; protein; 15–25 per cent; mineral matter, 1–2 per cent; and other constituents, 1 per cent. Water content is known to vary inversely with fat content.

Need for artificial recruitment has arisen in order to replace or augment stocks decimated by:

- Decline of water quality and destructive fishing (*e.g.,* pollution, poisoning, dynamiting);
- Barrier to migration caused by execution of river valley projects (*e.g.,* anadromous fish) and overfishing.

From the global view point, the fish which have overwhelmingly dominated artificial recruitment are: i) Oncorhynchus ii) Acipenser iii) Salmo. Artificial recruitment of carp, tilapia and mullet are also important mostly in tropical and subtropical countries. Oncorhychus and Salmo transplants have contributed maximum to sport and recreational fishing.

Production of livebait *e.g.,* for skipjack tuna (Katsuwonus pelamis) is an example of bait production for commercial fishing. Some potential live-bait species are: Tilapia mossambica, Dorosoma petenense, Engraulis japonicus, Sardinella malanure, several species of mullets and cyprinids.

A wide variety of ornamental fish such as sword tail (Xiphophrus helleri); angel fish (Pterophyllum scale), siamese fighter (Betta splendens), goldfish, and common carp. The last mentioned supports intensive breeding of fancy carps (live jewels) of Japan. There has come into being fish-cum-livestock culture, in the form of an integrated system especially involving cattle, pigs, ducks and poultry.

Several by-products are obtained from fish. They include fish meal used for animal feeding (in aquaculture an important component of most fish feeds) and as manure; fish flour; fish oil; leather; gelatin and glue from fish skins; imitation pearls; isinglass; adhesives; insulin from fish pancreas; sex hormones from gonads etc. Production of industrial fish includes production for purposes of reduction to fishmeal or fertilizers. Seaweeds are cultured for marine colloids and pearl oysters for cultured pearls.

STATUS OF AQUACULTURE

Compared with fishing, aquaculture is currently of little commercial significance to the Pacific Islands, with one important exception, black pearl farming, which is virtually confined to eastern Polynesia. Elsewhere in the Pacific, considerable development is needed before aquaculture can be considered economically sustainable. Shrimp (Penaeus spp.) farming has been a focus of commercial development in several islands with varying degrees of success; tilapia (Oreochromis niloticus) aquaculture has entered the subsistence economy in some areas, and seaweed (Kappaphycus spp.) is a future commercial export prospect. The culture of other marine and freshwater species is, however, generally still at the experimental or "backyard" stage.

The expansion of aquaculture in the Pacific will depend on providing better production methods for species currently being farmed, and techniques for propagating and growing the "new" species described above. These methods and techniques should be simple and flexible so that they can be adapted to the context of the Pacific Islands environment and to the market constraints (local and export markets). This approach should favour systems integrating fisheries and mariculture with low investment and operating costs and simple technical production processes. This should be done in association with pilot commercial-scale operations to test and demonstrate the economic viability of the methods proposed. This will require research combined with assistance, training and education programmes. Pacific Island nations have many attributes that favour development of aquaculture and stock enhancement in the coastal zone. These are as follows: a great diversity of coral reef species which are in high demand, proximity to major aquaculture and seafood markets in Asia, availability of suitable growout sites in pristine habitats, geographic conditions which favour restocking and stock enhancement, a relatively inexpensive labour force, and a tradition of working with marine resources.

Although these confer many advantages on the region in terms of aquaculture development and stock enhancement, there are still several constraints for such enterprises in the Pacific, which include limited domestic markets, high added-value export markets targeted, transport problems, socio-economic factors, fragile habitats, limited fresh water, and cyclones. Some of the best opportunities for aquaculture development in the Pacific are in the aquarium trade (coral reef fish, hard corals, soft corals), the live seafood markets (*e.g.*, groupers, spiny lobsters, abalone, crabs) and the pharmaceutical industry (*e.g.*, algae, sponges, soft corals). In all cases, the products are of high value and can be grown in small areas with relatively simple technology.

This review covers the insular Pacific as defined by the work-area of the Secretariat of the Pacific Community (SPC).

Compared to fishing, aquaculture is currently of little commercial significance in the insular Pacific, with one important exception-black pearl farming-and this is virtually confined to eastern Polynesia. Elsewhere in the Pacific, considerable development is needed before aquaculture can be considered economically sustainable. Shrimp (Penaeus spp.) farming has been a focus of commercial development in several islands over the past 30 years, with varying degrees of success; tilapia (Oreochromis niloticus) aquaculture has entered the subsistence economy in some areas, and seaweed (Kappaphyceus spp.) is considered a future commercial export prospect by the region. The culture of other marine and freshwater species is generally, however, still at the experimental or "backyard" stage.

Aquaculture is a relatively new development in the region, and in most Pacific Islands where it has been attempted; its history goes back less than 30 years. There is no fund of traditional knowledge for culturing fish and shellfish, just catching them, except in very specialized instances and areas. There is thus no great resource of aquacultural skill or infrastructure. This steep development path has perhaps not been taken into account in some development projects, which have often had unrealistic short-term aims and lacked follow-up.

Despite the comparatively minor penetration of aquaculture into Pacific Island economies, and despite the loss of interest by most of the international development community after many short-term project failures, several Pacific Island governments have accepted the challenge. They recognize that expansion in capture fisheries is limited, and have made substantial investments in freshwater aquaculture and mariculture, often in concert with external sources of development assistance.

Pacific Islanders cannot turn away from the sea. It is the greatest resource they have. The Potential Importance of Aquaculture to Pacific Islands: A major problem facing most of the island nations in the Pacific is that they have relatively few opportunities to generate income. The economies of most Pacific countries are limited due to small landmasses, few terrestrial resources and low numbers of inhabitants.

To ensure further development, island nations must make the most of the one important resource they all have-the sea. Through the joint efforts of the Forum Fisheries Agency (FFA) and the SPC Oceanic Fisheries Programme, island nations are deriving major inputs to their economies by fishing for tuna, or by selling access rights to tuna, within the large maritime zones under their control. However, valuable, sustainable harvests are also possible from inshore waters and coral reefs.

The potential for increased well-being of coastal communities from the responsibleuse of their inshore marine resources arises because the inshore habitats surrounding Pacific nations support a great diversity of economically important species (Wright and Hill, 1993; Dalzell *et al.*, 1996; Bell and Gervis, 1999). Traditionally, these animals were harvested at subsistence levels. More recently, however, development of export markets has provided coastal communities with opportunities to earn income from the inshore fisheries species.

Unfortunately, the transition from a subsistence to a market economy has usually been far from ideal: chronic overfishing has occurred in some areas. In such places, there are now too few of the prized animals to sustain reasonable harvests. Destructive fishing methods have compounded the problem by degrading some habitats to the point where they cannot support the valuable species. Pacific Island countries now recognize that aquaculture provides one of the few long-term, sustainable, ways of deriving benefits from inshore fisheries resources. This view of aquaculture as a priority area for continued sustainable development was reinforced as part of a consensus member country statement arising from the 2nd SPC Fisheries Management Workshop in 1998:

"...Regional fisheries managers have focussed on establishing regimes to sustain inshore fisheries. This is supported by a strategy to divert demand and fishing pressure to alternative activities, mostly to offshore fishing and into aquaculture.

Development of fisheries management varies from country to country reflecting the differing stages of economic development and levels of need. In some countries the need is to encourage economic activities and to generate income for rural villages; in other countries the need is to restrict or limit fishing. Yet in other countries the need to involve all stakeholders in the management system has evolved into community-based decision-making and control. There is now broad acceptance that marine resources cannot be managed in isolation from other users, or by one government agency so that an integrated and co-ordinated approach should be taken. In many circumstances, because of the smallness of the islands, an island system-management approach is the desirable option.

Aquaculture, as an alternative activity, is still at a preliminary stage of economic development in most PIC3, but is of enormous future significance.

For aquaculture to realise its full potential to the economies of PIC in a sustainable way will require a considerable degree of international support. PIC have endorsed a strategy to harness and prioritise such support at the regional institutional level. Several PIC already devote significant national resources to this subsector and this trend will continue as benefits are realised..." Advantages of Pacific Islands for Aquaculture: Pacific Island nations have many attributes that favour development of aquaculture and stock enhancement in the coastal zone. These include:

- A great diversity of species associated with coral reefs that are in high demand for:
- the aquaculture and seafood markets in Asia (*e.g.*, napoleon wrasse, groupers, sea cucumbers, spiny lobsters, trochus, pearl oysters, giant clams, green snail);
- the marine aquarium trade (*e.g.*, clownfish, angelfish, hard corals, soft corals, giant clams); and
- the pharmaceutical trade (*e.g.*, algae, sponges, soft corals, sea horses).
- Proximity to the major aquaculture and seafood markets of Asia-flight times are short enough to ensure that many species can be shipped alive to Asia.
- Availability of suitable growout sites in pristine habitats-coral reef lagoons create the calm conditions essential for culture of many species. The favourable environmental conditions should be the opportunity to develop green label products to get better prices on the international market.
- Geography that favours restocking and stock enhancement-most Pacific countries are small islands, or groups of small islands, surrounded by deep water. Cultured juveniles released into the inshore waters of island ecosystems cannot emigrate, and are therefore relatively easy to recapture.
- A relatively inexpensive labour force-expectations for financial return on labour are low in many Pacific countries relative to developed countries.
- A tradition of working with marine resources-coastal communities are already familiar with the basic biology of many species.

ORGANIC WASTE UTILIZATION IN AQUACULTURE

Organic waste utilization in aquaculture can either be extensive, with wastes occurring naturally or being added, with little or no further management of production, or it can be highly intensive under conditions where extraneous feeding and fertilization with inorganic plant nutrients play a major role in augmenting animal protein yield from the system. Aquatic animal husbandry is pursued in embayments, ponds, rivers, lakes, raceways, in brackish water, fresh water, and full-salinity ocean water. Many variants of extensive and intensive fish culture rely importantly on the growing together of a few compatible species (polyculture) to make fullest use of the various types of food present (plankton, bottom fauna) in a body of natural or managed water. Feeds and/or inorganic or organic fertilizers may also be added. Shellfish - molluscs and prawns or shrimp - can also be grown in polyculture, but managerial practices involving several species of fish exclusively are more advanced than those that combine the husbandry of invertebrates and fishes.

The rationale of polyculture is obviously to divert as much as possible of the attainable biomass into channels that are useful to man, compared to those that prevail in a wholly natural food web. It should be pointed out that stocking fish in a lake does not necessarily increase the productivity of the lake, only the productivity of harvestable (*i.e.*, desirable) organisms.

In a lake supporting a complex food web ending in only a few harvestable fishes, and many other non-desirable organisms (for human consumption), stocking with desirable fish fingerlings, as well as control of undesirable species, especially predators, diverts the biomass of the "non-suitable" species into production of desirable fish biomass. It should be added, though, that monoculture, especially of molluscs in brackish water and of predatory fish such as trout, salmon, or groupers, can furnish high annual yields that are commercially attractive in spite of substantial inputs: for these species, the food is extraneously supplied.

A brief comparison of materials flow in ponds versus cage culture seems appropriate here. The food webs of manured, fed, and fertilized polyculture ponds are exceedingly complex in comparison with cage monoculture systems.

The principal pathways of nitrogen in ponds. Possible nitrogen inputs to intensively managed polyculture ponds may be many and varied: atmospheric nitrogen, and nitrogenous substances present in in-flowing water, in inorganic fertilizers, in organic wastes, and in feeds. Similar diversity is also present in losses of nitrogen from ponds. Measurement of nitrogen flow between the various nitrogen pools (*e.g.*, dissolved nitrogen pool, bacterial nitrogen pool, etc.) is a difficult task, requiring sophisticated techniques.

For the present, aquaculturists are utilizing information on a few such inputs, pathways, and pools, and are only very slowly evolving models allowing prediction of harvests of fish from a pond given a specified set of conditions.

In contrast, cage culture represents a simplified ecosystem, at least within the confines of the cage itself. Cages are the aquatic counterpart to high-density terrestrial husbandry systems, including cattle feedlots, chicken batteries, and so forth. Such production systems are fashioned to provide all of the environmental needs of the animal and maximize production per unit surface area. The cage system is quite simple, with feed or food supplied to the fish (usually monoculture, but with polyculture cage culture becoming more common), and faeces, uneaten food, and waste metabolites being swept away from the cage by the continuous flow of fresh water.

Cage culture, however, is generally more capital-intensive than pond culture, and requires a considerably higher degree of knowledge of the nutritional needs of the fish, as the aquaculturist often supplies the fish with their sole source of food or feed.

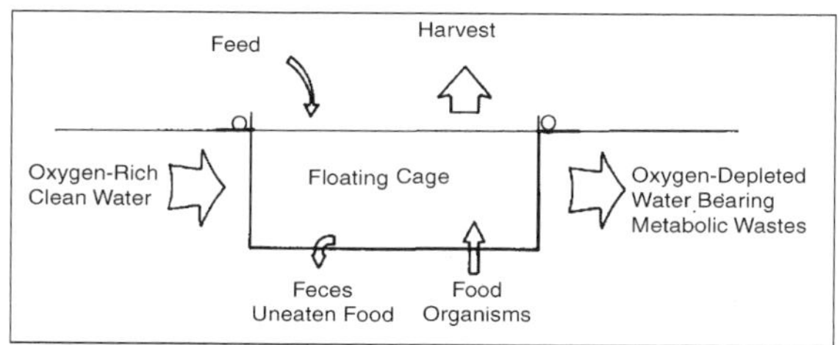

Fig. Generalized Flow of Materials Important in Cage Culture of Aquacultural Animals

STOCKING RATE IN FISH CULTURE

The denser the stocking rate in fish culture, the more difficult the management, largely because of the accumulation of waste substances, depletion of dissolved oxygen, and other problems of sanitation. Yet it is known from the interaction of sewage with rivers that fertilization does not necessarily lead to oxygen depletion when the river flow is reasonably swift, causing diffusion of atmospheric oxygen into the water.

Thus, it is not surprising that higher aquatic production is more often registered in flowing than in still waters, and that estuarine waters with their river-borne nutrients and continuous water circulation (via river flow and tides) represent the world's most fertile and productive aquatic environments. A comparison of annual productivity for several aquatic ecosystems is provided in Table, based on Crisp. Several compilations in the literature 11, 3 - 6) permit quantitative comparisons of these various conditions under more or less intensive management regimes. The time base of a hundred days rather than a year is chosen, on the advice of Ohla of the Hungarian Aquaculture Research Institute, so that summer production in the temperate zone can be compared with the average in the tropics.

The animal protein yield to man in natural waters of the temperate zone ranges from less than 20 to several hundred kg/ha. There are specific sites, often with inadvertent fertilization by run-off, where yields are much higher: a dead river arm in Hungary with a slow flow that also serves a duck farm produces 1.3 tons/ha/100 days, and a portion of the lake Laguna del Bay in the Philippines, the shallow recipient of much agriculture) and domestic drainage where the wind also concentrates the plankton, is capable of producing 3 tons or more of milkfish per ha/100 days; however, the site is now plagued with pollution from highly populated shores. A comparable level of fish productivity prevails in the more typical fish ponds of Israel or India, albeit with more material inputs.

Animal wastes are applied and often there is extraneous feeding; there also, the 100-day yield reaches 3 to 4 tons/ha. Sewage oxidation ponds that are stocked with polyculture species able to make the best use of the rapid production of algae and invertebrate biota yield between 2.4 and 4 tons of fish per ha/100 days.

Still higher yields are reached in flowing waters: Indonesian carp cages in sewage-fed streams produce 8 kg/m^2/100 days (cage surface area only), amounting to a practical harvest of at least 20 tons/ha in the same period. The apparent discrepancy is due to the fact that only a portion of the flowage can be obstructed with cages lest inundation become severe. Intensive feeding of the carp is not practiced because blood-worms thrive in the sediments of these streams, feeding on the bacteria-rich silt and ooze, and these worms are grazed upon by the carp. The peculiar aspect of the situation, however, is that the worms regenerate fast and easily after the fish nibble off a portion of them. The worms, incidentally, contain growth stimulators, probably hemoglobin-related compounds.

The relationship between fish stocking, density, and water flow rate. Although these values are species-specific, the benefits of increased oxygenation and removal of toxic body metabolites by rapid water exchange hold true for most species. Sewage oxidation ponds are also good bases for aquaculture. J. Ohla (personal communication, 1978) established that 2.5 tons/ha/100 days of severe) species of carp, mainly silver carp, can be grown if 200 m^3/ha/day of primary sewage effluent is added to the water (after settling of the solids).

The stocking of fish in sewage oxidation ponds is beneficial not only to fish culture, but to the waste treatment processes within the oxidation pond itself. Tilapia and silver carp were stocked in an oxidation pond and compared to an oxidation pond not containing fish. Bacteria levels were lower in the pond containing fish, perhaps due to the disinfection potential of waters with high pH and oxygen. The high pH probably results in a greater loss rate of ammonia to the atmosphere, considered to be beneficial from the waste-treatment viewpoints, and also from the viewpoint of fish health management (ammonia is toxic to fish). However, it represents a loss of valuable fixed nitrogen fertilizer, and hence can be a somewhat negative trade-off.

The use of raw or primary treated sewage has been questioned for health reasons. It should be noted here that intestinal parasites and flora of man and other warm-blooded animals are anaerobic, or nearly so, and that the high oxygenation of fast flowages, of balanced sewage ponds, or of balanced fish ponds receiving organic wastes, does not permit the survival of most such pathogens. Finally, thorough cooking provides an additional safeguard.

The addition of sewage to pond or river water, and the use of fertilization by means of animal wastes in aquaculture, are well established and economically sound practices. Higher yields per unit surface area are reached with other extraneous inputs, mainly by supplementary feeding and/or by forced water circulation or even filtering. The main secret of success in achieving up to several hundred tons/ha annual fish production (often extrapolated from smaller surface areas) lies in the use of natural or artificial flowages in which many thousands of fish can be stocked and fed. Similarly, cage-reared fish in strong tidal flows can achieve phenomenal rates of growth and production. They are fed with compounded feed (however expensive), with trash fish, with household wastes, or with cereals, cereal wastes, or other agricultural residues. Attractive economic returns conditioned by cultural food tastes invite these practices.

Yet another high-yield aquatic production system ought to be mentioned, this time in the sea. In certain bays near Vigo, Spain, where tidal exchange as well as fertile run-off from the land are high, mussels are grown on containment/attachment devices suspended from rafts. Three hundred tons of mussel flesh per year have been reported. Under roughly comparable conditions, similar planktonfiltering bivalves (Mytilussmaraydinus) also reach such high yields (10). Tidal movement-based mollusc culture, however, is more advanced in Europe and Japan than it is in tropical Asia; non-etheless, in all locations it is threatened by pollution.

Having noted the beneficial effects of dissolved organic wastes on aquacultural systems, one must hasten to add that most fish farmers use chemical fertilizers, predominantly or entirely. Manure is just not sufficiently ubiquitous, and the use of domestic sewage, even while more widely available, has strong cultural and economic barriers against its use. While the cultural component of this barrier is perhaps understandable, though ill-advised, the economic one invites further comment.

Where sanitation is advanced, investments in rendering sewage effluents innocuous and infertile are so great that sewage aquaculture can only be envisaged when cost of invested capita) need not be considered. That is, in the temperate zone and in technologically advanced countries, one may think of planning sewage aquaculture for small to medium-sized towns with readily available cheap land or a nearby lake. Installations will be necessary to separate domestic sewage from all other liquid effluents such as industrial wastes and non-sewage domestic and stream run off components. Even then, big-accumulation of toxic substances may occur. The actual levels of trace metal accumulation will be species- and site-specific. For example, rainbow trout fed a diet containing activated sludge selectively accumulate certain metals, but not others.

In the tropics, as already indicated with reference to the cage culture for carp in Indonesia, the situation may be more favourable to the direct use of domestic sewage because there is, as yet, less admixture with the effluents of industrial or hydrocarbon chemicals. Aquaculture and agriculture in mainland China utilize both animal and human wastes. Reports suggest that, while animal wastes are normally either applied untreated, or composted and applied, anaerobic treatment and composting are the preferred methods for human wastes. This process greatly reduces parasites and pathogens, rendering the manure safer to use.

WASTEWATER USE IN AQUACULTURE

FOOD CHAINS

The objective in fertilizing an aquaculture pond with excreta, nightsoil or wastewater is to produce natural food for fish. Since several species of fish feed directly on faecal solids, use of raw sewage or fresh nightsoil as influent to fish ponds should be prohibited for health reasons.

Edwards has represented the complex food chains in an excreta-fed fish pond, involving ultimate decomposers or bacteria, phytoplankton, zooplankton and invertebrate detritivores. Inorganic nutrients released in the bacterial degredation of organic solids in sewage, nightsoil or excreta are taken up by phytoplankton. Zooplankton graze phytoplankton and small detritus particles coated with bacteria, the latter also serving as food for benthic invertebrate detritivores.

Plankton, particularly phytoplankaton, are the major sources of natural food in a fish pond but benthic invertebrates, mainly chironomids, also serve as fish food, although they are quantitatively less important. To optimize fish production in a human waste fed pond, the majority of the fish should be filter feeders, to exploit the plankton growth.

A wide range of fish species has been cultivated in aquaculture ponds receiving human waste, including common carp *(Cyprinùs carpio),* Indian major carps (*Catla catlax, Cirrhina mrigala and Labeo rohita),* Chinese silver carp *(Hypophthalmichthys molitrix),* bighead carp *(Aristichthys nobilis),* grass carp *(Ctenopharyngodon idella),* crucian carp (*Carassius auratus),* Nile carp *(Osteochilus hasseltii),* tilapia *(Oreochromis spp.),* milkfish *(Chanos chanos),* catfish *(Pangasius spp.),* kissing gouramy *(Helostoma temmincki),* giant gourami *(Osphronemus goramy),* silver barb *(Puntius gonionotus)* and freshwater prawn *(Macrobrachium lanchesterii).*

FISH SPECIES

The selection reflects local culture rather than fish optimally-suited to such environments. For example, Chinese carps and Indian major carps are the major species in excreta-fed systems in China and India, respectively. In some countries, a polyculture of several fish species is used. Tilapia are generally cultured to a lesser extent than carps in excreta-fed systems although, technically, they are more suitable for this environment because they are better able to tolerate adverse environmental conditions than carp species. Milkfish have been found to have poorer growth and survival statistics compared with Indian major carps and Chinese carps in ponds fed with stabilization pond effluent in India. Edwards gives a thorough review of current knowledge on the various fish species which can be cultured in ponds fed with human waste. It would appear that considerable confusion still exists with regard to fish feeding on natural food. Although fish are generally divided into types according to their natural nutritional habits - those that feed on phytoplankton, or zooplankton or benthic animals - several species are known to feed on whatever particles are suspended in the water.

There is also uncertainty about the types of phytoplankton fed upon by filter-feeding fish. For example, although blue-green algae are thought to be indigestible to fish, Tilapia have been shown to readily digest these algae and there is evidence that silver carp can do the same.

AQUATIC PLANTS

Aquatic macrophytes grow readily in ponds fed with human waste and their use in wastewater treatment. Some creeping aquatic macrophytes are cultivated as vegetables for human consumption in aquaculture ponds and duckweeds are also cultivated, mainly for fish feed. Among the aquatic plants grown for use as vegetables are water spinach *(Ipomoea aquatica),* water mimosa (*Neptunia oleracea),* water cress *(Rorippa nasturtium-aquaticum)* and Chinese water chestnut *(Eleocharis dulcis).* The duckweeds *Lemna, Spirodela* and *Wolffia* are cultivated in some parts of Asia in shallow ponds fertilized with excreta, mainly as feed for Chinese carps but also for chickens, ducks and edible snails.

AQUACULTURE TECHNOLOGY

The technology employed in aquaculture has developed over many centuries but has done so more rapidly during the last half century. Aquaculture systems, and the technology used, vary from very simple systems, used for family ponds in tropical countries, where production is for domestic consumption, to high technology systems, such as intensive closed systems, like those used for rearing stripped bass. Herbivorous and filter feeding fish, reared mostly in simple systems of small freshwater ponds, however, account for about half of global aquaculture production.

Much of the technology used in aquaculture is relatively simple, amounting to small modifications that improve the growth and survival rates of the target species, such as providing additional food, adding seed animals collected elsewhere, managing water exchange to maintain adequate oxygen levels, and protecting the stock from predators. Greater understanding of complex interactions between nutrients, bacteria and cultured organisms, together with advances in hydrodynamics applied to pond and tank design, have enabled the development of closed systems. These have the advantage of isolating the aquaculture systems from natural aquatic systems, thus minimizing the risk of disease or genetic impacts on the external systems.

Developments in engineering, some learnt from offshore oilrig construction, increase the possibilities for offshore aquaculture using robust cages. Sea ranching, the release of young fish into the wild to improve the harvest in capture fisheries, has also made a start but its long term viability is still to be assessed. Major advances are also being made in the technology of the production of aquafeeds, which generally require the combining of a large number of ingredients into very small feed pellets.

AQUACULTURE ENGINEERING

Aquaculture draws on well-established engineering fields for most of the design and construction needs of its production facilities. Building earth ponds is similar to building roads - a knowledge of the characteristics of soils and the limits of safe design are the basis of good construction. Similarly, the buildings used for hatcheries and other support activities are no different from those common in the housing, agricultural and commercial sectors. Sometimes ponds are lined with plastics or other impermeable materials, and here the techniques are similar to those for civil structures such as potable water reservoirs or sludge tanks.

The design and installation of water control gates, including in unstable soils, benefits from the long experience in this field in the agriculture and irrigation sectors. An understanding of hydrodynamics allows ponds and tanks to be built with good water circulation, oxygen mixing and without 'dead spots' where sediments might accumulate and cause health problems to fish.

For installations in the sea, the situation is somewhat different and many of the important engineering solutions, such as for fish cages or suspended shellfish growout systems, have had to be developed by aquaculturists themselves. They have benefited however from the accumulated knowledge of seafarers in general and fishermen in particular, in the design and operation of mooring and buoyage systems.

More recently, when fish farmers have turned their attention to how to operate fish cages in locations further offshore where seas are rougher, the experience of the oil exploration industry has proven very valuable. Techniques have been developed in recent years for the production of fish and other aquatic products in closed recirculation systems. To make these work, aquaculturists have needed to develop a knowledge of the biological processes operating - such as how bacteria can be used to neutralise and re-cycle the nitrogenous waste products produced by growing fish - and how to engineer the systems to meet the biological requirements. Knowledge of bio-engineering from the waste treatment and water treatment industries has made contributions to the development of closed aquaculture systems and there is probably more that could be usefully transferred from the sewage treatment sector to help solve problems in fish rearing.

Modern materials have brought many benefits to aquaculture. For instance, custom made plastic joints have simplified the construction of sea cages, and made them more reliable in high stress conditions. Experiments with huge free-floating or sunken net cages operated in the open ocean were begun several decades ago, for instance in the Caspian Sea. These showed some promise, but more reliable construction materials will make the farming of fish in such structures increasingly feasible.

Modern materials and production methods have been important also, in the construction of plastic filter substrates for indoor recirculating systems. The fine detail of these has been found to make substantial differences to the efficiency of biological filters.

In shrimp farms, specially designed matting materials that stand upright on pond bottoms, with a structure that promotes the growth of the small animals and plants that the shrimp can thrive on, have recently been developed and shown to boost production.

Engineering skills are important in the design of most aquaculture facilities and good engineering can affect the efficiency and economics of production. If capital costs can be minimised while still maximising productivity and reducing risk, the farming operation will be more profitable. Aquaculturists have proven very innovative over the past fifty years, constantly developing new technologies to support their farming operations. As new production methods and species for farming are developed, the engineering solutions needed to support them will continue to evolve.

Aquaculture Facilities

Aquaculture began with man making small modifications to natural habitats so as to improve the survival and growth of target species. Some of the oldest examples are in the rearing of freshwater fish in ponds, which has been practiced for thousands of years in Asia and at least for many centuries in Europe. The simple act of placing a mesh barrier across the outlet of a small pond or lake to prevent fish from escaping can make a big improvement in the supply of food. Similarly, along coasts, some tidal lagoons can be easily turned into ponds. Closing off such naturally occurring water bodies was the start, centuries ago, of much fish and shrimp aquaculture in Asia and, in modern times, in South America.

Removing predators and improving the conditions within the pond, (for instance by providing more area of preferred water depth), supplying additional food and, later, by adding seed animals collected outside, were further steps that moved aquaculture production close to where it is today. The farming of seaweeds and molluscs (oysters, clams, mussels etc) developed similarly, as people made improvements such as providing more settlement areas for the young, or the removal of predators from the growing area.

Farmers had a natural desire to improve the productivity of their systems and, as knowledge grew, they learned to stock more animals, increase feeding and manage the exchange of water to maintain the conditions, such as adequate oxygen levels, that the animals needed to survive. The range of facilities used for aquaculture subsequently broadened for a number of reasons. Farmers encountered difficulties with more intensive rearing because of the uncontrolled influence of pond soils, local water quality, weather. Some of these problems could be resolved by rearing in ponds built of concrete, or lined with plastic, by bringing the ponds indoors under cover, or treating the water before flowing it to the culture ponds.

Secondly, as the naturally occurring ponds and lagoons all became used, prospective farmers had to take a broader approach and develop the technology and engineering to be able to use less naturally-favored sites. Net cages floating in protected coastal or inland waters were developed for fish culture, or fish were stocked in fenced areas of the sea or of large lakes. At the same time, the need for supplies of young animals (fry, seed) to stock these systems led to the development of hatchery techniques and dedicated hatchery facilities. For most species this aspect of production has proved more successful when conditions can be more closely controlled, for instance in indoor concrete and fibreglass tank systems, rather than in outdoor ponds. Often, the younger stages of aquatic animals are more sensitive than adults to physical and chemical conditions and these have to be managed within a smaller range, if production is to be successful. Thus hatchery facilities developed as a separate branch of the industry.

More recently, our knowledge has improved greatly regarding the complex interactions that occur in a rearing system, between nutrients, bacteria and the cultured organism. This and technological developments have allowed many aquatic organisms to be reared in completely closed recirculating facilities, including the farming of marine organisms at locations far from the sea. Closed systems have the added advantage of offering greater protection from the danger of disease entering from the natural environment and also of minimizing adverse effects of the production system on that external environment. Technology has also begun to open up the possibilities of growing fish in enclosures in the open ocean, something that could one day transform the nature of human food production on the planet. With 70 per cent of the earth's surface covered by water, the potential is clear. Earlier technology has restricted the cage farming of fish to sheltered coastal waters. Here the number of available sites is limited, environmental damage is more likely and conflicts exist with other users. As cages are developed that can withstand the demanding conditions of the open ocean, or which can be operated below the ocean surface, farming could move further offshore.

A small start has been made, notably in Japan, through so-called 'ranching' programmes', the farming of the sea without enclosures where fry are released in large numbers into the ocean with the aim of improving returns from the capture fishery. Sometimes artificial reef structures are created underwater as well, to increase the available habitat and natural food for the fish.

Aquaculture Systems

Aquaculture - the growing of aquatic animals and plants - covers a wide range of species and methods. Some of the simplest production systems are the small family ponds in tropical countries where carp are reared for domestic consumption. At the other end of the scale are high technology systems, such as the intensive indoor closed units used in the USA for the rearing of striped bass or the sea cages used in Chile and Europe for growing salmon and bream. Nearly half of the world's aquaculture however is of herbivorous and filter-feeding fish such as carp.

Practised in freshwater, in ponds made of earth, and often managed by a single household, production systems are simple and in many cases have changed little over many centuries. A farmer will excavate a small pond near his house; sometimes, as for instance in the flood river deltas of countries like Vietnam, using the spoil to build a raised foundation for the house itself. The pond will typically be fed by natural rainfall or groundwater, or sometimes by diverting a nearby stream or irrigation canal. Some stocking occurs naturally, but more commonly a farmer will buy young fish from a breeder and stock them in his pond, often after holding them for a time in a small floating net known as a 'hapa' to check that they are in good condition before release in the pond. Production will improve if the farmer can stock a good balance of different species to make the best use of the varied kinds of food available in the pond. Typically the fish will be fed with household or agricultural by-products, and harvested when the family needs a meal.

Stocking and harvesting is often continuous, where knowledge of the best stocking level is learned from experience and where the pond is not drained for many years. For most such farmers, fish production is a secondary activity, a useful additional source of protein to add to the supply or income from his main agricultural or commercial activities.

Outdoor freshwater fish farming is practised commercially in many countries, including in the developed nations. The density of stocking may be much higher, the control of feeding, water quality and fish health more closely monitored, but even in intensively operated systems, the key parameters that need to be monitored and balanced for success are similar to those in the traditional household pond system.

Over the last half-century, systems for the indoor rearing of many species of fish have also been developed. Holding of fish in controlled conditions indoors has been important in the development of seed production (hatcheries). Also, as knowledge has increased of nutrient cycles, bacterial action and water chemistry, it has become possible to rear fish, both for food and for ornament, in 'closed' indoor systems, where the water is recirculated. Passing the water through filters that use bacteria to naturally break down and recycle the waste products produced by the fish allows a production system to be run in almost total isolation from the outside. Such systems have been important in the control of disease and also in maintaining the stable conditions that some species need to flourish and reproduce.

A fifth of current world aquaculture production is of plants - mainly seaweeds, that for the most part are grown on the seabed or on raft or racks in shallow coastal waters and used directly for food or for the production of alginate or carageenan (agar-agar). This sector of aquaculture is another largely run by small-scale growers, mainly in Asia but in some parts of the South America, who attach seedlings of marine plants to simple structures built close to shore and constructed from natural materials and then harvest the plants once they have grown. A further fifth of world production is of molluscs such as oysters, clams and mussels.

Again the majority of production is small-scale and by coastal people who often combine fishing activities with farming. Oysters and mussels are mainly grown on structures built above the seabed - poles or racks on the shore, or ropes suspended from rafts or floating lines. The farmer's role is to supply suitable places for seed to settle or to add to natural production by bringing seed from a hatchery - and then to maintain the conditions of waterflow and freedom from predators that the shellfish need to grow. Most commercially grown molluscs feed on microscopic algae floating in the water, so the farmer does not need to provide any feed.

Fish are also grown in coastal areas, in ponds or in floating cages. As technology advances, there is the potential to develop systems for rearing fish in the open ocean, either in sturdy cages or by so called 'ranching', where young fish are released to the wild and then collected by normal fishing or by training them to respond to specially generated sounds.

The tropical coastal zones of Asia and Latin America are where most shrimp farms are found. Shrimp farming is mainly carried out in earth-bottomed ponds built on flat land close to the sea. Shrimp are raised in ponds at a range of different densities. At low densities the systems need only limited inputs of feed and fertilizers. As the density increases, more feed has to be supplied and at the higher end of the range, so-called 'intensive' farming, machines have to be put in the ponds to mix and aerate the water.

Aquaculture is practised from the cold waters of the far north and south, where fish like salmon, arctic char and sturgeon are grown in ponds, flowing raceways and cages in the sea, down through the latitudes to the tropics, where carp and tilapia flourish in freshwater and shrimp and seabass are farmed along the coasts. It ranges from production of fish in naturally occurring ponds in rural areas to the intensive culture of ornamental fish in plastic tanks in the middle of a city. It is practised by the poorest farmers in developing countries as a livelihood and to supply much needed protein for their families - and by urban sports shop owners in Europe and the U.S.A. producing baitfish for weekend anglers. Systems can range from an intensive indoor system monitored with high-tech equipment through to the simple release of baby fish to the sea - but all with the same aim, of helping boost Nature's natural productivity

Biotechnology Development

Biotechnology - any technological application that uses biological systems, living organisms, or derivatives thereof, to make or modify products or processes for specific use 1994 UNEP Convention on Biological Diversity - has a wide range of useful applications in fisheries and aquaculture. It creates opportunities, for instance, to increase growth rate in farmed species, boost the nutritional value of aquafeeds, improve fish health, help restore and protect environments, extend the range of aquatic species and improve the management and conservation of wild stocks.

Some biotechnologies are simple with a long history of application such as the fertilization of ponds to increase feed availability. Others are more advanced and take advantage of increasing knowledge of molecular biology and genetics, *e.g.*, genetic engineering and DNA disease diagnosis.The field of genetic biotechnology similarly ranges from simple techniques such as hybridization, to more complex processes such as the transfer of specific genes between species to create GMOs (genetically modified organisms). Over the years, our knowledge of fish breeding requirements has improved and the ability to induce artificial breeding developed through the use of natural or synthetic hormones and/or environmental manipulations. (For example changing photoperiod or water temperature can induce some fish to spawn). These have been key factors facilitating the application of more advanced biotechnologies.

Selective breeding, the maintenance of stocks genetically improved by chromosome manipulation, line crossing, and sex reversal all depend on the controlled breeding of farmed species. These improvements in reproductive technologies have also assisted aquaculturists greatly in their efforts to domesticate aquatic animals. In addition, by making it possible to remove the natural constraints and timing of breeding, farmers are able to mate many more species at times that are most beneficial, and thus helping to ensure a steady and consistent supply of fish for consumption.

IMPORTANCE OF FEEDS FOR THE FUTURE OF AQUACULTURE

Fish meal has been the protein source of choice in diets for farmed fish for several reasons. First, the protein content of fish meals is relatively high, generally 65-72 per cent, depending upon the fish species used to produce the fish meal. Second, the amino acid profile of fish meal closely matches the dietary requirements of most carnivorous fish species. Third, protein and amino acid apparent digestibility is relatively high in good quality fish meal in most farmed fish species.

Finally, fish meal-based diets are highly palatable to most farmed fish. Of these properties of fish meal, plant protein sources are similar with respect to apparent protein and amino acid digestibility, and protein concentrates are similar to fish meal in protein content. However, amino acid profiles of plant protein sources do not match the dietary requirements of carnivorous fish species as well as do the amino acid profiles of marine proteins, and some plant protein sources lower feed intake, presumably by lowering feed palatability, when replacement levels are high. Thus, the role of plant proteins up to now has primarily been to replace a portion of the fish meal protein in diets for carnivorous fish species to lower the price of the feed, extend fish meal supplies when they are tight, or to reduce the total phosphorus level of the diet. Certain protein ingredients from the rendering industry, *e.g.*, blood meal and feather meal, have been used in a similar fashion, and also to increase the total protein level of the diet.

The amount of protein supplied by fish meal in diet formulations for various species of fish differs significantly, depending on whether or not the species is carnivorous or omnivorous. Salmon and trout, for example, are fed diets that contain 38-44 per cent crude protein during the grow-out stage, where most feed is used during a production cycle.[59] Catfish, in contrast, are fed diets containing 28-32 per cent crude protein, most of which is supplied by soybean meal. Members of the carp family are fed diets with protein contents varying from 0 to 35 per cent, depending on species, where they are farmed and life-history stage. Fry and fingerling carp are fed diets containing higher protein levels than are post-juvenile fish.

Carp diets intended for use in high-input rearing systems contain 15-25 per cent fish meal, and although this is a relatively low fish meal inclusion level, the tremendous increase in high-input carp culture has dramatically increased the amount of fish meal used by this production sector to about 17 per cent of the total amount of fish meal used in all aquaculture diets in 2000. Together, 2.1 million tons of fish meal was used in diets for fish and shrimp in 2000. The percentage of fish meal in the diet of various species groups ranged from 55 per cent for marine flatfish (flounder, turbot, halibut) to 3 per cent for catfish (channel catfish, African catfish). Carp averaged 5 per cent, but this figure includes both high-input and low-input systems. Carp farming is converting to high-input systems, and this will increase the total use of fish meal in this production sector, despite an anticipated reduction in the percentage of fish meal used in diets. Carp feed production is anticipated to increase from about 7 million tons in 2000 to 27 million tons by 2010. Soybean meal will likely supply the bulk of protein in carp diets of the future, but fish meal will continue to be used, especially in diets for fry and fingerling carp.

AQUACULTURE FEED PRODUCTION

Barlow predicted that aquaculture feed production would increase from 13.1 million tons in 2000 to 37.2 million tons in 2010. At today's fish meal use levels in diet formulations for various species groups, the amount of fish meal needed to produce 37.2 million tons of fish and shrimp feed would slightly exceed the amount of fish meal traded worldwide in non-El Nino years. However, Barlow[15] predicts that the percentage of fish meal in diet formulations will decrease and that total fish meal use by the aquaculture industry will be 2.8 million tons of fish meal in 2010. Total protein in these diet formulations will not change.

The difference between 4.6 million tons and 2.8 million tons, *e.g.*, ~1.7 million tons, will be supplied by other protein sources.

Assuming that the fish meal used in fish feeds contains 70 per cent crude protein, then ~1.2 million tons of protein from sources other than fish meal will be needed annually in fish feeds by 2010. If soybean meal (48 per cent crude protein) were used to supply this protein, the increase in total use in all aquaculture feeds would be ~2.6 million tons. If soy protein concentrate were used, the total would be less, approximately 1.6 million tons because of its higher protein content. If other protein concentrates from grains or oilseeds, *e.g.*, wheat gluten meal, corn gluten meal, canola protein concentrate, were used, the amounts would be similar.

Eleven "Great Trends" Identified by the SPARE Panel

1. We live in a closely connected world in the midst of a communications revolution, where an acceleration of the pace of change is occurring.
2. We live on human-dominated marine and inland coasts; these coastal societies are removing invaluable ecosystems and habitats at unprecedented rates and delivering never-before seen levels of pollution to aquatic ecosystems. These major "drivers of change" are putting into question the future survival of the world's traditional, working coastal societies.
3. Urbanization and agriculture are causing increasing water shortages in many societies. Water withdrawals from freshwater and estuarine ecosystems are having large-scale impacts on aquatic ecosystems.
4. Because of the shift to the coast, the coherency of individual stakeholder groups is disappearing. Instead of a "fishing community" we have multiple users struggling for survival, with many groups shifting around between sectors and geographic regions. Large-scale seasonal migrations between inland and coastal ecosystems and cities now occur regularly. Our attention to and understanding of this "shifting nature of survival" is very poor.

5. We live in a transitional time in fisheries where economic failures and huge subsidies in fishing are occurring. However, this situation is changing rapidly. We expect the next big international battlefront will be fishing subsidies: which countries are handing them out; and what should be done about them.
6. Commercial fishing and global fishing fleets will continue to shrink. The silver lining is that we might end up with a much more efficient and financially sound fishing fleet in 20 years.
7. There is a dramatic globalization and consolidation of seafood trade and labour, with a major shift of seafood production to developing countries. The longer the world economy remains sluggish; the more consolidation we are likely to see.
8. Continual failures to manage capture fisheries rationally have led to a global aquaculture industry growing more rapidly than any other resource-based sector. Aquaculture will grow at an even quicker pace over the next 20 years, then slow due to environmental constraints.
9. Concerns over the safety and purity of the food we put in our mouths is fueling a new era of food safety. Expect more food safety regulations in the future. Expect environmental labeling and other market-based approaches to ensuring sustainable fisheries and aquaculture to become commonplace worldwide.
10. There has been a spectacular rise in the power and scientific capabilities of international NGOs. While NGOs have filed lawsuits and started boycotts, expect them to challenge the university and government science and management establishments more regularly. One reason for this is the terrible government neglect of applied fisheries and aquaculture science and management institutions in government and universities worldwide.
11. Lastly, in the midst of these breathtaking changes and extraordinary challenges, the "sustainability imperative" has arisen like a phoenix. But to date, governments and sustainability visionaries have given little attention to the dire needs of the living resources of the world's oceans and freshwaters.

Lessons from Agroecology

There are many opportunities for aquaculture and fisheries to employ farming systems research and extension methods derived from agroecology to aquaculture farming and fishing systems in developing countries.

By a recent estimate, some 4.4 million farmers applying agroecological methods have achieved productive results, mostly in marginal environments. Farmers using agroecological models have achieved significant levels of production security while also achieving natural resource conservation. Farmers employ methods such as polycropping, crop-livestock integration, agroforestry, integrated soil management and integrated pest management and have been shown capable of producing substantial and secure yield increases.

Conversion to agroecological methods requires more labour, significantly greater contributions of farmers' knowledge, and greater "adaptive abilities". After the ecological infrastructure is established, however, agroecosystems are able to sustain their own plant protection, soil fertility, and farm productivity.

Agroecology consists of a universal set of principles, which translate into site-specific technological forms according to local needs and socio-economic circumstances. Farmers' knowledge and experimentation are essential resources. New strategies focus on the facilitation of farmer learning to become experts in natural resource management, and the involvement directly in the formulation of the research agenda, with active participation in the process of technological innovation and outreach/dissemination. In addition, researchers focus their attention on translation of general ecological principles of natural resource management into practical advice directly relevant to smallholder farmers.

Even though there is now greater interest in diverse, low input systems, agroecological systems are not regarded by many agricultural scientists as capable of meeting the production needs of the future. As a result, research on agroecology has been largely ignored by mainstream agricultural research, which has continued to pursue conventional approaches towards yield maximization of specific crops with large external inputs, and biotechnology.

Comparative Advantages of the USA in Fisheries and Aquaculture

The USA possesses many comparative advantages over other nations in a number of disciplines, techniques and technologies. These advantages should be applied in a systematic way to resolve issues in fisheries and aquaculture among developing countries. Here we use comparative advantages related to geographical information systems (GIS), remote sensing, mapping and GPS to illustrate the issues, spatial solutions, comparative advantages and the strategies that should be considered for implementation. The strategies are general, and many comparative advantages in other fields could be listed.

Aquaculture has grown at a rapid pace, especially in recent years. Nevertheless, unplanned growth, or growth that was poorly planned, has resulted in a number of social, economic, environmental and administrative problems. Likewise, coastal and inland food fisheries continue to be threatened by degradation of the environment and overfishing. The confluence of these problems makes the sustainability of aquaculture and inland fisheries the central issue. All of the main issues in aquaculture and inland fisheries have spatial elements to some degree, and several of the issues quite explicitly call for spatial solutions that can be most effectively addressed by the deployment of GIS, remote sensing and mapping individually, or in concert. For example, fisheries and aquaculture problems that have been addressed by GIS, remote sensing and mapping include: (1) Zoning for fish reserves, and for aquaculture development; (2) Identification of essential fish habitat; (3) Aquaculture information systems for planning and management of aquaculture among government agencies; (4) Assessing the potential for aquaculture development (suitability for systems and productivity of organisms); (5) Integrating fisheries and aquaculture into coastal area management, and river and lake basin management; (6) Assessing the status of fishery resources and of aquatic habitats; (7) Estimating environmental impacts on fisheries and aquaculture, and planning for mitigation; (8) Estimating the effects of fisheries and aquaculture on the environment; and (9) Aquaculture and fisheries early warning (*e.g.*, harmful algal blooms, lethal temperatures). A salient problem is that, with few exceptions, developing countries that contribute most importantly to global aquaculture and inland fisheries production appear not to be deploying GIS, remote sensing and mapping tools to improve the sustainability of these sectors.

The USA has a marked advantage in spatial analyses. In the realm of aquaculture, the USA accounted for 36 per cent of all global applications of GIS from 1985 to 2002. The situation is even more remarkable with respect to inland fisheries: the USA accounted for 71 per cent of the world total. Experience in these applications comes from a variety of sources: US government, state and local agencies, universities, and NGOs.

The USA is the world leader in the development and marketing of GIS software. In a related field, the US dominates in inexpensive global positioning system (GPS) units that have become indispensable for incorporating ground and aquatic field data into maps and GIS. More generally, the USA is the preeminent provider of inexpensive computers, peripherals and software, including Internet servers. The US government operates a range of satellite-borne sensors, some of which are especially designed to collect data from aquatic environments. The coverage is frequent and world wide. Historical data are archived making change analyses possible. These data are the potential backbone of much of the spatial analyses needed to address aquaculture and fisheries problems among developing countries.

The goal is to improve the sustainability of aquaculture and inland fisheries through practical, needs-driven training on applications of GIS, remote sensing and mapping. The following strategies are advocated:

- Place emphasis on achieving self-reliance in spatial analyses for fisheries and aquaculture in a measured, step-wise fashion with solutions appropriate to capabilities and sustainability, and frequent evaluation of results.
- Provide a range of services according to assessed needs and capabilities to sustain the initiatives provided via USAID. For example,

because the statistical data are so incomplete, all fisheries and aquaculture organizations can benefit from mapping, but few countries in Africa could support a GIS aimed at fisheries and aquaculture alone.

- Concentrate on practical training in-country, and on fellowships that provide a wide range of theoretical and practical training in the USA.
- Distance learning via the Internet, TV and radio is increasingly being used for training in developing countries. Even though the US appears to possess the means to be effective in these media, training on fisheries and aquaculture using these media is not yet common. This opportunity should be evaluated.
- Work with many organizations according to their comparative advantages: national and sub-national fisheries and aquaculture government entities, commercial firms, trade organizations, universities and professional organizations.
- In developing nations, work outside of fishery and aquaculture sectors with entities that have broader or overlapping responsibilities with the goal to integrate aquaculture and fisheries into broader GIS-based planning initiatives and to mobilize to issues of common concerns for the environment.
- A "Fisheries and Aquaculture International Training Coordination Unit" could be established in USAID to identify the training needs and opportunities, to contract with the providers and to evaluate the results of the training.

Fisheries are in the news. A flurry of media activity has centered around fisheries issues in the past year prompted by the release of several studies and reports that point to growing crises and controversy in both wild fisheries and aquaculture. A recent report from a panel of fishermen, scientists, business leaders, and government officials pointed to overfished and depleted stocks in U.S. waters, along with severe habitat degradation. The report argued that the restoration of U.S. fisheries requires a major overhaul of policy, including the introduction of ecosystem-based management and stronger regulations. A much-publicized study in *Nature* reported that the population of large predatory marine fish has been reduced by 90 percent since pre-industrial times. Another recent study argued that correcting reported Chinese fisheries statistics to levels that better fit estimates of biophysical potential renders global catch trends far less favorable (Watson and Pauly 2001). The Food and Agriculture Organization of the United Nations (FAO), particularly in its State of World Fisheries and Aquaculture publications, has consistently sounded the alarm over threatened stocks of wild fish.

The rapidly growing field of aquaculture, which now accounts for 30 percent of the world's food fish, has also pushed its way into the media spotlight. For some years now, aquaculture has been seen as a possible savior for the overburdened wild fisheries sector, and an important new source of food fish for the poor.

However, there are some problems with the industry. A recent report from the World Wildlife Fund argued that some forms of aquaculture place pressure on wild fisheries through demand for wild-caught fish as feed. Another Pew report warned of the lack of effective regulatory mechanisms for dealing with genetically modified fish, some varieties of which are already in development (Pew Initiative on Food and Biotechnology 2003).

Numerous studies have warned of potentially negative effects of escaped farmed fish on wild populations. A 2000 study in *Nature* argued that while aquaculture has the potential to contribute considerably to the world's supply of fish, significant environmental tradeoffs have occurred with many forms of aquaculture. In response, many industry advocates and policymakers have strongly defended aquaculture as an environmentally sustainable means of contributing to the world's fish supplies.

In this context, what does this study contribute to the growing dialogue on world fisheries? Why would economists attempt to address issues of fisheries that typically have been dealt with by biologists, ecologists, and policymakers? The answer: a key missing component to the puzzle thus far has been a broad economic analysis of the rapid changes in fisheries over the past two decades. Economic factors have been a crucial driver of these changes in fisheries, and economic factors will drive further changes to the year 2020. It is imperative that audiences beyond the fisheries sector—especially policymakers—have a better understanding of fisheries issues, and how they interact with other critical policy issues in world food and agriculture.

Most critical among the issues addressed in this study are those of poverty reduction and environmental sustainability in developing countries. The intention is to address the issues in a consistent economic framework that focuses on the dynamic and interacting decisions of producers, consumers, and traders all over the world, at the same time permitting sufficient data disaggregation by commodity and location to support useful conclusions on the likely future of fisheries.

A better understanding of these market interactions is not a substitute for other aspects of forecasting fisheries outcomes, such as stock assessment, fish population dynamics, and biophysical modeling; however, better information on price-mediated inter-relationships with supply and demand for other foods and feeds is essential for reasoned policymaking. We trust that this study addresses this need to some degree.

CRITICAL CHANGES IN THE FISHERIES SECTOR

The overall purpose of the study is to analyze the changing—and now critical— place of fisheries in global food policy issues. The term "fisheries" is applied equally to the capture of wild fish and to aquaculture. The focus is developing countries, although the analysis includes the developed world. In a sector as globalized as fisheries has become over the past two decades, events in one part of the world impinge quickly on outcomes in another.

The study starts from a series of premises that are documented and supported as the book unfolds. Attention is paid to critical changes in the fisheries sector and their impacts on a broader set of policy objectives for growth, poverty reduction, and environmental sustainability in developing countries.

Many past and future developments in the fisheries sector can best be understood by considering fish as a series of commodities within a changing world supply and demand system for different food and feed items. The study concludes with a delineation of key domains for policy action within the fisheries sector that can improve outcomes for broader food and agricultural development policy objectives in developing countries. Thoughts on priorities for further food policy research in the fisheries area are also provided.

The study investigates the following key premises:

- Fish production has developed from a primarily local and relatively minor specialized activity into an important part of a rapidly evolving global high-value food activity.
- Limits to wild capture fisheries mean that the time for fish farming has come—but how and with what perils remains to be seen.
- Capture fisheries affect aquaculture both as a competitor and as a supplier of feed, leading to a complicated price-mediated relationship that is generally not well understood.
- Poor rural people in developing countries are major stakeholders in the future of fisheries—and this is true, furthermore, for an expanding inland rural population.
- Tremendous uncertainties inherent in capture fisheries affect aquaculture in key ways, particularly through markets for inputs and outputs.
- Policies and technology development in both developing and developed countries will be crucial to improving global outcomes.

We investigate these premises first by focusing on what is changing and then by devising a way to project how these trends might play out under different scenarios. Historical consumption, production, price, and trade trends are assessed for a wide variety of regions and for a number of fisheries products. The study then attempts to illustrate the economic component that cuts across these important multidisciplinary issues and cannot be ignored. The analysis aims to link modeled outcomes to issues for environmental sustainability, technology generation, trade policy, poverty alleviation, and human nutrition in poor countries.

A key objective is to put the consideration of fisheries issues into the broader context of evolving world markets for food. Looking at fish as a series of market commodities with different market profiles draws explicit attention to the role of prices in affecting both consumer and producer behaviour towards different sub-categories of fish. It also permits examination of market tradeoffs within fisheries and between fish and other foods. Fish is both a food market competitor and an input to meat production, with nearly one-third of all wild-caught fish being used as a feed input. Even within fisheries, competition exists among different kinds of fish, and some kinds of fish provide critical inputs to the production of others. Events affecting prevailing prices for different kinds of fish and meat affect other commodities in the animal protein group. Furthermore, fish feed prices also potentially interact with prices for vegetable protein sources, such as soy.

9

Biological Crisis of Fishery Management

From this point, fishery sciences itself are progressively switching their attention from single species to ecosystem approach, from micro to macro perspectives, increasing the need for measuring the impact of fishing on natural and man-made systems. Consequently, Cochrane argued that the problems currently experienced in fishery management throughout the world occur in four realms, namely biological, ecological, economic and social crises. Biological crisis of fishery management started during the last decade after awareness grew about the alarming status of fishery resources. Ludwig for example, drew attention to collapses of fish stocks such as those of the Pacific salmon, the Californian sardine, and the Peruvian anchovy. In 1994, FAO showed, in an analysis, of global fish landings, that there had been reduction in the annual growth in landings in 1980s, and that in 1990 there had been 3 per cent reduction in the global annual catch from the previous year. Garcia and Newton analyzed that this trend continued over the next few years and in between 1990 and 1992, global landings fell by an average of 1.5 per cent per year.

In an economic perspective, fisheries actually exist to meet social and economic demands and one would expect to find that the impact on the resources has resulted in a measurable social and/or economic benefit. Unfortunately, some evidences suggest that the expected benefits have not been in the form of economic gain.

Christy estimated that the gross revenues from the total global marine landing in 1989 was US$ 70 billion, giving a deficit balance since the global operating costs were estimated at a level of US$ 92 billion. With an annual capital costs of US$ 32 billion per year, the total deficit in global fisheries was estimated to be US$ 54 billion per year.

At a country level, Japan for example, a deficit gross cash flow was also founded for cause of purse seine fisheries. In this type of fisheries, the volume of production was estimated to be lower than its MYS, OSY as well as MEY level. However, due to its high operating cost and low price of fish caught by the purse seine, it resulted in a deficit cash flow. Matsuda called this phenomenon the poverty in large catch. Nevertheless, the recent study undertaken by Tietze shows that in some fisheries, they perform a positive gross cash flow in their global operation. Out of 108 types of vessels from 15 countries throughout the world, 97 per cent had a positive gross cash flow and fully recovered their operating costs. When also considering the annual cost of capital, 85 per cent showed a net profit after deducting the cost of depreciation and interests. However, in terms of sustainability issues, some symptoms of economic crisis in some fisheries should be considered in order to come up with the solution for economic sustainability in the system.

From a social point of view, fisheries are rarely seen as simple tools for generating economic returns. The role of fisheries as a source of employment, particularly in rural or more remote areas, has also widely been given high priority. Moreover, McGoodwin elaborated that fisheries actually are a human phenomenon which essentially are places where human activities are linked with marine ecosystems and renewable resources. Human fishing activities defined the attribute of fishery, since without it there would only be an aquatic realm where various marine species live. Therefore, it has been strongly argued that clearly the fisheries are much more than geographic regions, fishing methods, types of fishing gear, fish species, or economic-domains— it is something more human.

SOCIAL ASPECTS IN FISHERIES

In this regard, social aspects in fisheries would be very important to be understood. One of the major features of fisheries in recent decades, related to the social crisis, has been the introduction of modern fishing technologies and also the increasing globalization of trade affecting the fishing communities. Modernization in fisheries has two faces of which actually one contributed to the welfare of fishing communities, while the other has led to the social problem related to the depletion of fishery resources. According to Crean and Symes, in the North Atlantic and Mediterranean, there has been a decline in the quality of life and standard of living among many fishing communities, as also experienced in some parts of South East Asia. This symptom of social crisis in fisheries should seriously be taken up when we consider the sustainability in this sector. We argue that the need for measuring and evaluating the sustainability of fishery activities in a system perspective has acquired its highest importance and should be undertaken at various levels involving all aspects in the fishery system.

As also strongly argued by Charles that critical concern about sustainability arises not only in terms of catch level or even biomass level, but in all aspects of the fishery: from the ecosystem, to the social and economic structure, to the fishing communities and management institutions, as well as fish stocks themselves. Moreover, the pursuit of sustainable fisheries is best seen as requiring, more than keeping the catch of fish, to keep a level that is not too large.

Instead, sustainability can be viewed comprehensively as the maintaining or enhancing of four key components, namely ecological, socioeconomic, community and institutional sustainability indicators. In Chuenpagdee and Alder, there have been some methods for analyzing and evaluating fishery sustainability, such as RapFish method, FAO Code of Conduct Compliance, and International Instrument Compliance.

RapFish is a rapid appraisal technique designed to allow an objective, transparent, multi-disciplinary evaluation. It uses multidimensional scaling, and ordination method to appraise the relative sustainability of fisheries. RapFish relies on a defined-scoring of a number of attributes in five dimensions, namely of ecological, economic, social, technological and ethical. As its characteristics of a rapid appraisal, this method has a slightly top-down approach by employing the defined-score and criteria for the fishery system evaluation. The application of RapFish can be seen, for example, in the Gulf of Maine fisheries, in the North Atlantic Fisheries and recently in the case of Jakarta coastal fisheries, Indonesia.

Similar to RapFish, compliance with the FAO Code of Conduct required multidimensional scaling approach. In this case, four different sets of attributes articulate the clauses in the Code. This evaluation method reflects national management intentions and management practices for each fishery system that is assessed. The application of this method has been carried out in Australian fisheries, Australian multi-species trawl fishery, and Gulf of Maine fisheries. International Instrument Compliance method focuses on measuring either qualitatively or quantitatively the level of compliance with the fisheries management provisions contained within an instrument. In this method, criterion scores ranged from 0 (low compliance) to 3 (high compliance) and the total criteria were limited to approximately six per instrument. An example of this method is presented by Chuenpagdee and Alder in the case of Gulf of Maine fisheries.

It could be said that the above methods rather focus on relatively static approach from the top (top down approach). The fishery in question is just used like the static object which will be evaluated using such methods. They do not involve stakeholders to evaluate their fishery. By these methods, a straight forward analysis could be done but should be treated as preliminary results only. We then argue that it is necessary to confirm dynamic complex systems of the fishery by using a bottom-up approach.

This study attempts to evaluate sustainability indicators of a fishery system by involving the local stakeholders rather than by straightforward analysis (top down) as shown in other measurement methods previously described. We assess what we call ¯local accepted fishery sustainability by examining local sustainability indicators, their importance, interactions, connections, linkages, and assess sustainable state indicators using a cognitive approach involving fishery stakeholders in the area studied. Indicators of sustainability in this case could be defined as a variable that can be used to measure the status or condition of a system.

Information on cumulative impacts of indicators is important because those indicators are often linked together and consequently may impact sustainability directly and indirectly. It is also argued that although individual indicators appear to be good, fishery sustainability may be compromised due to cross-criterion or cross-indicators interactions. Further, the overall risk of a fishery system is not only affected by the risk associated with each indicator, but also by the ecosystem-wide risk that emerges from cross-indicator linkages. Therefore, while being an indicator in and by itself may not be a risk but it may eventually be compromised because of its linkages either directly or indirectly to other indicators that are at risk.

In this study, we used the fishery system of Yoron Island, Kagoshima Prefecture, Japan as the case study. This island is located in the southernmost of Kagoshima Prefecture which is bordered by Okinawa Prefecture in the south, Okinoerabu Island in the north, South China Sea in the west, and Pacific Ocean in the east. Geographically, this island is included in the group of Amami Islands together with Okinerabu Island, Amami Oshima, Tanegashima, and Kikai Island. Among these islands, Yoron Island is the smallest and in terms of area as is estimated to be 20.5 and 23.7 km of coastal line. In terms of population, this island is also small with only 6210 people inhabiting in year 2000. This island is also considered to be relatively vulnerable in terms of its economy according to Adrianto and Matsuda.

Fishery sector in Yoron Island was considered as one of the main areas of economic activities compared to other islands in the group of Amami Islands. Using location analysis approach, fishery sector in Yoron Island has been considered to have very high representation of its economy with location quotient (LQ) value of 1.82, compared with the LQ values of other islands that have lower than 1.0 except Amami Oshima. In terms of production, during the 10 years of analysis, fishery sector in Yoron Island was considerably stable and tended to be growing. In 1988, the fishery production was 321 ton and increased to 384 ton in 1998. In 2002, the total production was recorded to be 540 ton.

Beyond the statistic performance of fishery production, it was important to have an in-depth analysis on the importance of sustainability issues in fishery development in the island. First, as previously elaborated, fishery sector has been an important sector in Yoron Island, so that such efforts to maintain the performance of fisheries in this island are considerably important in terms of its economic function to increase the gross island products. Secondly, it is a matter of fact that fishery sector in Yoron has an important role in the food supply of the islanders. In 2003, almost a half of the fishery production was consumed locally. Out of the 540.139 ton total production, 139.638 ton went to local market as food supply for the local islanders. From these two reasons, we may argue that analysis of sustainable fishery system in Yoron Island is considerably needed from ecological, socioeconomical, and institutional point of views.

THE BACTERIAL FLORA ON LIVE FISH

Microorganisms are found on all the outer surfaces (skin and gills) and in the intestines of live and newly caught fish. The total number of organisms vary enormously and Liston (1980) states a normal range of 10^2–10^7 cfu (colony forming units)/cm^2 on the skin surface. The gills and the intestines both contain between 10^3 and 10^9 cfu/g.

The bacterial flora on newly-caught fish depends on the environment in which it is caught rather than on the fish species. Fish caught in very cold, clean waters carry the lower numbers whereas fish caught in warm waters have slightly higher counts. Very high numbers, *i.e.*, 10^7 cfu/cm^2 are found on fish from polluted warm waters. Many different bacterial species can be found on the fish surfaces. The bacteria on temperate water fish are all classified according to their growth temperature range as either psychrotrophs or psychrophiles. Psychrotrophs (cold-tolerant) are bacteria capable of growth at 0°C but with optimum around 25°C.

Psychrophiles (cold-loving) are bacteria with maximum growth temperature around 20°C and optimum temperature at 15°C. In warmer waters, higher numbers of mesophiles can be isolated.

The microflora on temperate water fish is dominated by psychrotrophic Gram-negative rodshaped bacteria belonging to the genera *Pseudomonas, Moraxella, Acinetobacter, Shewanella* and *Flavobacterium.*

Members of the *Vibrionaceae (Vibrio* and *Photobacterium)* and the *Aeromonadaceae (Aeromonas* spp.) are also common aquatic bacteria and typical of the fish flora. Gram-positive organisms as *Bacillus, Micrococcus, Clostridium, Lactobacillus* and coryneforms can also be found in varying proportions, but in general, Gram-negative bacteria dominate the microflora. Shewan (1977) concluded that Gram-positive *Bacillus* and Micrococcus dominate on fish from tropical waters.

However, this conclusion has later been challenged by several studies which have found that the microflora on tropical fish species is very similar to the flora on temperate species. A microflora consisting of *Pseudomonas, Acinetobacter, Moraxella* and *Vibrio* has been found on newly-caught fish in several Indian studies. Several authors conclude, as Liston (1980), that the microflora on tropical fish often carry a slightly higher load of Gram-positives and enteric bacteria but otherwise is similar to the flora on temperate-water fish.

Aeromonas spp. are typical of freshwater fish, whereas a number of bacteria require sodium for growth and are thus typical of marine waters. These include *Vibrio*, *Photobacterium* and *Shewanella*. However, although *Shewanella putrefaciens* is characterized as sodium-requiring, strains of *S. putrefaciens* can also be isolated from freshwater environments. Although *S. putrefaciens* has been isolated from tropical freshwaters, it is not important in the spoilage of freshwater fish.

Table: Bacterial Flora on Fish Caught in Clean, Unpolluted Waters.

Gram-negative	Gram-positive	Comments
Pseudomonas	*Bacillus*	
Moraxella	*Clostridium*	
Acinetobacter	*Micrococcus*	
Shewanella putrefaciens	*Lactobacillus*	
Flavobacterium	Coryneforms	
Cytophaga		
Vibrio Photobacterium		*Vibrio* and *Photobacterium*
Aeromonas		are typical of marine waters; *Aeromonas* is typical of freshwater

In polluted waters, high numbers of *Enterobacteriaceae* may be found. In clean temperate waters, these organisms disappear rapidly, but it has been shown that *Escherichia coli* and *Salmonella* can survive for very long periods in tropical waters and once introduced may almost become indigenous to the environment.

The taxonomy of S. *putrefaciens* has been rather confused. The organism was originally associated with the *Achromobacter* group but was later placed in the Shewan *Pseudomonas* group IV.

Based on percentage of guanine+ cytosine (GC per cent) it was transferred to the genus *Alteromonas,* but on the basis of 5SRNA homology it was reclassified to a new genus, *Shewanella*.

It has recently been suggested that the genus *Aeromonas* spp. which was a member of the *Vibrionaceae* family be transferred to its own family, the *Aeromonadaceae*. Japanese studies have shown very high numbers of microorganisms in the gastrointestinal tract of fish, and as such numbers are much higher than in the surrounding water, this indicates the presence of a favourable ecological niche for the microorganisms.

Similarly, Larsen *et al.* (1978) reported up to 10^7 cfu/g of vibrio-like organisms in the intestinal tract of cod and Westerdahl *et al.* (1991) also isolated high numbers of vibrio-like organisms from the intestines of turbot. *Photobacterium phosphoreum* which can be isolated from the surface can also be isolated in high numbers from the intestinal tract of some fish species. On the contrary, some authors believe that the microflora of the gastrointestinal tract is merely a reflection of the environment and the food intake.

Microbial Invasion

The flesh of healthy live or newly-caught fish is sterile as the immune system of the fish prevents the bacteria from growing in the flesh. When the fish dies, the immune system collapses and bacteria are allowed to proliferate freely. On the skin surface, the bacteria to a large extent colonize the scale pockets. During storage, they invade the flesh by moving between the muscle fibres. Murray and Shewan (1979) found that only a very limited number of bacteria invaded the flesh during iced storage.

Ruskol and Bendsen (1992) showed that bacteria can be detected by microscope in the flesh when the number of organisms on the skin surface increases above 10^6 cfu/cm^2.

This was seen at both iced and ambient temperatures. No difference was found in the invasive patterns of specific spoilage bacteria (*e.g.*, S. *putrefaciens)* and non-spoilage bacteria.

Since only a limited number of organisms actually invade the flesh and microbial growth mainly takes place at the surface, spoilage is probably to a large extent a consequence of bacterial enzymes diffusing into the flesh and nutrients diffusing to the outside.

Fish spoil at very different rates, and differences in surface properties of fish have been proposed to explain this. Skins of fish have very different textures. Thus whiting *(Merlangius merlangus)* and cod *(Gadus morhua)* which have a very fragile integument spoil rapidly compared to several flatfish such as plaice that has a very robust dermis and epidermis. Furthermore, the latter group has a very thick slime layer, which includes several antibacterial components, such as antibodies, complement and bacteriolytic enzymes.

THE WORKING PRINCIPLES OF FISHERIES MANAGEMENT

A complex and possibly confusing picture of all the tasks which need to be considered by the fisheries manager. Some of this complexity can be reduced by attempting to highlight the underlying key issues. There are both benefits and risks in attempting to simplify a subject and over-simplification can lead to neglect of important details. However, simplification can facilitate understanding important principles and highlighting the broad areas which need attention. Arising from the considerations, a number of key principles can be identified which may serve to focus attention on the starting points for effective fisheries management.

WHO IS THE FISHERY MANAGER

The Technical Guidelines (FAO, 1997) suggest that fisheries management institutions have two major components: the fisheries management authority and the interested parties. The fishers and fishing companies would usually be the major participants amongst the interested parties. The fisheries management authority is that entity which has been given the mandate by the State (or States in the case of an international authority) to perform specific management functions.

In many countries that authority would be a Department of Fisheries or, within a broader Department, a Division of Fisheries. However, a fisheries management authority does not have to fall directly within central government, and could be, for example, provincial, local, parastatal or private. Any one of these arrangements can function effectively, given an adequate legal framework in which to operate and the resources necessary to fulfil their function.

Who, then, within this authority is the fisheries manager and to whom is this Guidebook addressed? In fact, despite the fact that we have deliberately used the term in the title, we suggest that in modern fisheries management, there is rarely a single individual who fulfils the functions of "fisheries manager". The head of the authority, for example, a Director of Fisheries, may have overall responsibility for implementing fisheries management and, as well as being accountable and responsible for the advice passed on from his or her Department to the political decision-maker, may act in an overall coordinating role. However, this individual is unlikely to, and generally should not, have sole responsibility for receiving information, formulating advice and making and implementing decisions.

Fisheries management is a complex and multi-faceted discipline and requires input from a range of perspectives. It is therefore inappropriate to expect any individual to fulfil this function on their own. Fisheries management should involve the legitimate interested parties in the management process.

Perhaps the closest we can come to a "fisheries manager" is the management authority as a whole, including technical experts, monitoring, control and surveillance (MCS) units, administrative units, the executive body of the formal authority, the consultative mechanisms, the advisory body where one exists, and the responsible political head who is often a Minister. Each member of these functional bodies is, to some extent, a fishery manager and this Guidebook is aimed at all of them.

It is not designed to go into great technical and operational detail on each function or task as this would require a set of Guidebooks. Instead, it is intended to give a holistic picture of how the different functions should interact in a fisheries management authority in order to develop appropriate objectives, management strategies and plans, and how to encourage all those participating in a fishery to collaborate in and adhere to the agreed strategy.

CONSTITUTES A MANAGEMENT AUTHORITY

The responsibility for fisheries management rests with the designated fisheries arrangement or organisation which, in this Guidebook, we have referred to without distinction as the fisheries management authority. Following the practice used in the Technical Guidelines on Fisheries Management (FAO, 1997) the term is used broadly here to describe that legal entity which has been designated by the State as having the mandate to perform specified fisheries management functions. In practice, it may be a national or provincial ministry, a department within a ministry, or an agency and could be governmental, parastatal or private. In the case of shared resources it should be international.

The area of competence, geographical area, fish resources and fisheries for which a given management authority is responsible must be precisely specified in each case in the appropriate legislation. The task of an authority is diverse and complex and as a result, fisheries management authorities are normally divided into institutional support structures: the fisheries management institutions.

The institutions need to encompass the basic tasks and functions of fisheries management. The actual institutional structure and mechanisms may differ from authority to authority and it would be inappropriate for us in this Guidebook to attempt to prescribe any specific set of characteristics as representing the 'best' institutional structure and processes. What is best in each case will depend in large part on the specific circumstances and context. What is universal, however, is that it is essential for the different institutions concerned with management of any fishery or fisheries to be able to interact effectively, requiring good channels of communication and feedback. The institutions must also be seen by the different interested parties as being legitimate.

The need for collaboration between the authority and the interested parties is as important as collaboration between the institutions within the authority. Examines the pre-requisites for effective partnerships between the management authority and the interested parties and the different types of partnership which can be considered.

It is common and frequently desirable for the national government to devolve all or some fisheries management functions to local government or to smaller groups such as fishing communities. In such cases, it is essential to specify precisely the responsibilities and functions, including the geographical area, falling under this local authority or smaller group. The institutions within the local authority must follow the same principles as those for a national authority.

The Code of Conduct requires that fisheries management should be concerned with the whole stock over its entire area of distribution and therefore that States should cooperate in the management of transboundary, straddling, highly migratory and high seas fish stocks exploited by two or more states.

General rules for cooperation towards conservation of such fish stocks are foreseen in the United Nations Convention on the Law of the Sea of 10

December 1982, and in the 1995 UN Fish Stocks Agreement. The responsibilities, functions and structure of international or regional fisheries authorities will usually not differ substantively from those of national authorities.

GOALS AND OBJECTIVES

The over-riding goal of fisheries management is the long-term sustainable use of the fisheries resources. Achieving this requires a proactive approach and should involve actively seeking ways to optimise the benefits derived from the resources available.

This rarely happens, though, and fisheries management is still most commonly practised as a reactive activity, where decisions are made and actions taken largely in response to problems or crises. The resulting crisis decisions are then normally attempts merely to solve the immediate problems without properly considering the broader perspective and the longer-term objectives. Such an approach may succeed in maintaining dissatisfaction sufficiently low to avoid major conflict, but it is extremely unlikely to result in the best use of the marine resources being exploited by the fishery.

The first step in proactive fisheries management is to decide what is meant by optimising the benefits for each fishery - what can the State or the collection of legitimate interested parties agree on as being optimal benefits? This may be described in general terms in the national fisheries policy which must be the starting point for determining the specific objectives for each fishery. The broad goals stated in the fisheries policy may need to be tailored for a specific fishery, but the goals for each fishery should be consistent with the policy.

In general terms, the goals in fisheries management can be divided into four subsets: biological; ecological; economic and social, where social includes political and cultural goals. The biological and ecological goals may be more correctly thought of as constraints in achieving desired economic and social benefits but for simplicity and consistency with the terminology most commonly used in fisheries management, we will include them as goals in this Guidebook. Examples of goals under each of these categories include:

- To maintain the target species at or above the levels necessary to ensure their continued productivity (biological);
- To minimize the impacts of fishing on the physical environment and on non-target (bycatch), associated and dependent species (ecological);
- To maximize the net incomes of the participating fishers (economic); and
- To maximize employment opportunities for those dependent on the fishery for their livelihoods (social).

Identifying such goals is important in clarifying how the fish resources are to be used to benefit society, and they should be agreed upon and recorded, both at the policy level and for each fishery. Without such goals, there is no guidance on how the fishery should be operated, which results in a high probability of *ad hoc* decisions and sub-optimal use of the resources (resulting in lost benefits), and increases the probability of serious conflicts as different interest groups jostle for greater shares of the benefits. This is often seen in practice and one of the important causes of failures in fisheries management has been identified as the frequent absence of clear and precise objectives.

While setting goals is an essential first step, the goals stated above have two obvious limitations. Firstly, they have clear conflicts in intention as it is impossible, for example, to minimize impacts of the fishery on the ecosystem and simultaneously to maximize net incomes. Similarly, it is very probable that management strategies that aim to maximize net incomes will not also maximize employment opportunities. Some compromise between these goals has to be achieved before an effective management strategy can be devised. The second limitation of the goals is that they are too vague to be of much benefit to the manager.

For example, the impacts of fishing can only be "minimized" by having no fishing at all, which is unlikely to have been the intention of those who stated the goal. Maximising employment opportunities could mean allowing as many fishers as possible to participate, regardless of whether or not they could make a living from the fishery, or it could mean maximising the number which could still earn some acceptable income, or many other such targets. Too much is left to the discretion of the manager with these examples of goals.

It is therefore necessary to refine the goals further and to develop operational objectives for each fishery. Operational objectives are very precise and are formulated in such a way that they should be simultaneously achievable in that fishery. In other words, the trade-offs between the biological, ecological, economic and social goals must have been agreed upon and the conflicts and contradictions resolved. The development of operational objectives is illustrate the difference between goals and operational objectives, two examples of objectives are:

- To maintain the stock at all times above 50 per cent of its mean unexploited level (biological);
- To maintain all non-target, associated and dependent species above 50 per cent of their mean biomass levels in the absence of fishing activities (ecological).

With operational objectives such as these, it is possible for any observer, including the manager, to establish whether or not they are being achieved and hence whether or not the management strategy is appropriate and being successfully implemented. These operational objectives can also easily be used as the foundation for reference points, which are essentially the operational objectives expressed in a way which can be estimated or simulated in a fisheries assessment. Once operational objectives have been agreed upon, a management strategy can be developed, made up of a suite of different management measures, to achieve those objectives.

All of this may sound complex, but in reality is no more than most people do in order to develop a budget for their personal finances. Most of us have realistic but imprecisely expressed hopes and needs for our lifestyle as well as a knowledge of the nature of the resource (in this case our net income). These hopes and needs are the goals of our budget but they will all compete for the same resource, our net income, so there will probably be conflicts which need to be resolved.

Therefore we have to modify our goals and express them more precisely: we develop operational objectives in which we specify what we can realistically achieve in terms of food, housing, education etc. Thereafter, we need to decide on our budget strategy: how can we meet those objectives: what type and quantities of food and clothing should we be buying; what type of housing can we consider; can we consider an annual holiday, etc.

Clearly, our operational objectives must be consistent with the yield we can expect from the resource (our income). Normally the process of developing realistic objectives will require trade-offs and most of us find, for example, that we cannot allocate as much for entertainment or holidays as we would like and at the same time make our rental or mortgage payments. Therefore priorities are established and compromises made until eventually we arrive at realistic objectives that balance our desires with our income, and that provide a good guide on how to manage our finances from month to month and in the longer-term.

At the end of this, we should have a feasible financial management strategy that, barring totally unexpected events, will have a predictable outcome. If we have done our calculations correctly and responsibly, the strategy should mean we enjoy a reasonable lifestyle without being sued for bankruptcy. This is little different from the basic task, and overall hope, of the fisheries manager!

BIOLOGICAL ASPECTS AND CLASSIFICATION OF FISH

Fish are generally defined as aquatic vertebrates that use gills to obtain oxygen from water and have fins with variable number of skeletal elements called fin rays . Five vertebrate classes have species which could be called fish, but only two of these groups - the sharks and rays, and the bonyfish - are generally important and widely distributed in the aquatic environment. Fish are the most numerous of the vertebrates, with at least 20000 known species, and more than half (58 per cent) are found in the marine environment. They are most common in the warm and temperate waters of the continental shelves (some 8 000 species). In the cold polar waters about 1 100 species are found. In the oceanic pelagic environment well away from the effect of land, there are only some 225 species. Surprisingly, in the deeper mesopelagic zone of the pelagic environment (between 100 and 1 000 m depth) the number of species increases. There are some 1 000 species of so-called mid- water fish .

Classifying all these organisms into a system is not an easy task, but the taxonomist groups organisms into natural units that reflect evolutionary relationships. The smallest unit is the species. Each species is identified by a scientific name which has two parts the genus and the specific epithet (binominal nomenclature).

The genus name is always capitalized and both are italicized. As an example, the scientific (species) name of the common dolphin is *Delphinus delphis.* The genus is a category that contains one or more species, while the next step in the hierarchy is the family which may contain one or more genus. Thus the total hierarchical system is: Kingdom: Phylum: Class: Order: Family: Genus: Species.

The use of common or local names often creates confusion since the same species may have different names in different regions or, conversely, the same name is ascribed to several different species, sometimes with different technological properties. As a point of reference the scientific name should, therefore, be given in any kind of publication or report the first time a particular species is referred to by its common name. For further information see the International Council for the Exploration of the Sea "List of names of Fish and Shellfish" ; the "Multilingual Dictionary of Fish and Fish Products" prepared by the Organisation for Economic Cooperation and Development and the "Multilingual Illustrated Dictionary of Aquatic Animals and Plants" (Commission of the European Communities, 1993). The classification of fish into cartilaginous and bony (the jawless fish are of minor importance) is important from a practical viewpoint, since these groups of fish spoil differently and vary with regard to chemical composition.

ANATOMY AND PHYSIOLOGY

The Skeleton

Being vertebrates, fish have a vertebral column - the backbone - and a cranium covering the brain. The backbone runs from the head to the tail fin and is composed of segments (vertebrae). These vertebrae are extended dorsally to form neural spines, and in the trunk region they have lateral processes that bear ribs. The ribs are cartilaginous or bony structures in the connective tissue (myocommata) between the muscle segments (myotomes). Usually, there is also a corresponding number of false ribs or "pin bones" extending more or less horizontally into the muscle tissue. These bones cause a great deal of trouble when fish are being filleted or otherwise prepared for food.

Muscle Anatomy and Function

The anatomy of fish muscle is different from the anatomy of terrestrial mammals, in that the fish lacks the tendinous system connecting muscle bundles to the skeleton of the animal. Instead, fish has muscle cells running in parallel and connected to sheaths of connective tissue (myocommata), which are anchored to the skeleton and the skin. The bundles of parallel muscle cells are called myotomes.

All muscle cells extend the full length between two myocommata, and run parallel with the longitudinal direction of the fish. The muscle mass on each side of the fish makes up the fillet, of which the upper part is termed the dorsal muscle and the lower part the ventral muscle.

The fillet is heterogenous in that the length of the muscle cells vary from the head end (anterior) to the tail end (posterior). The longest muscle cells in cod are found at about the twelfth myotome counting from the head, with an average length around 10 mm in a fish that is 60 cm long . The diameter of the cells also vary, being widest in the ventral part of the fillet. The myocommata run in an oblique, almost "plow-like" pattern perpendicular to the long axis of the fish, from the skin to the spine. This anatomy is ideally suited for the flexing muscle movements necessary for propelling the fish through the water.

As in mammals, the muscle tissue of fish is composed of striated muscle. The functional unit, *i.e.*, the muscle cell, consists of sarcoplasma containing nuclei, glycogen grains, mitochondria, etc., and a number (up to 1 000) of myofibrils. The cell is surrounded by a sheath of connective tissue called the sarcolemma.

The myofibrils contain the contractile proteins, actin and myosin. These proteins or filaments are arranged in a characteristic alternating system making the muscle appear striated upon microscopic examination. Most fish muscle tissue is white but, depending on the species, many fish will have a certain amount of dark tissue of a brown or reddish colour. The dark muscle is located just under the skin along the side of the body.

The proportion of dark to light muscle varies with the activity of the fish. In pelagic fish, *i.e.*, species such as herring and mackerel which swim more or less continuously, up to 48 per cent of the body weight may consist of dark muscle . In demersal fish, *i.e.*, species which feed on the bottom and only move periodically, the amount of dark muscle is very small.

There are many differences in the chemical composition of the two muscle types, some of the more noteworthy being higher levels of lipids and myoglobin in the dark muscle.

From a technological point of view, the high lipid content of dark muscle is important because of problems with rancidity.

The reddish meat colour found in salmon and sea trout does not originate from myoglobin but is due to the red carotenoid, astaxanthin. The function of this pigment has not been clearly established, but it has been proposed that the carotenoid may play a role as an antioxidant. Further, the accumulation in the muscle may function as a depot for pigment needed at the time of spawning when the male develops a strong red colour in the skin and the female transport carotenoids into the eggs. The latter seems to depend heavily on the amount of carotenoids for proper development after fertilization. It is clearly seen that the muscle colour of salmonids fades at the time of spawning.

The fish cannot synthesize astaxanthin and is thus dependent on ingestion of the pigment through the feed. Some salmonids live in waters where the natural prey does not contain much carotenoid, *e.g.*, in the Baltic Sea, thus resulting in a muscle colour less red than salmonids from other waters. This may be taken as an indication that the proposed physiological function of astaxanthin in salmonids explained above may be less important.

In salmon aquaculture, astaxanthin is included in the feed, as the red colour of the flesh is one of the most important quality criteria for this species.

Muscle contraction starts when a nervous impulse sets off a release of Ca + + from the sarcoplasmic reticulum to the myofibrils. When the Ca + + concentration increases at the active enzyme site on the myosin filament, the enzyme ATP-ase is activated.

This ATP-ase splits the ATP found between the actin and myosin filaments, causing a release of energy. Most of this energy is used as contractile energy making the actin filaments slide in between the myosin filaments in a telescopic fashion, thereby contracting the muscle fibre. When the reaction is reversed (*i.e.*, when the Ca + + is pumped back, the contractile ATP-ase activity stops and the filaments are allowed to slip passively past each other), the muscle is relaxed.

The energy source for ATP generation in the light muscle is glycogen, whereas the dark muscle may also use lipids. A major difference is, further, that the dark muscle contains much more mitochondria than light muscle, thus enabling the dark muscle to operate an extensive aerobic energy metabolism resulting in CO_2 and H_2O as the end products. The light muscle, mostly generating energy by the anaerobic metabolism, accumulates lactic acid which has to be transported to the liver for further metabolization. In addition, the dark muscle is reported to possess functions similar to those are found in the liver.

The different metabolic patterns found in the two muscle types makes the light muscle excellently fitted for strong, short muscle bursts, whereas the dark muscle is designed for continual, although not so strong muscle movements. Post mortem the biochemical and physiological regulatory functions operating in vivo ceases, and the energy resources in the muscle are depleted. When the level of ATP reaches its minimum, myosin and actin are interconnected irreversibly, resulting in rigor mortis.

The Cardiovascular System

The cardiovascular system is of considerable interest to the fish technologist since it is important in some species to bleed the fish (*i.e.*, remove most of the blood) after capture.

The fish heart is constructed for single circulation. In bony fish it consists of two consecutive chambers pumping venous blood towards the gills via the ventral aorta.

After being aerated in the gills, the arterial blood is collected in the dorsal aorta running just beneath the vertebral column and from here it is dispersed into the different tissues via the capillaries. The venous blood returns to the heart, flowing in veins of increasingly larger size (the biggest is the dorsal vein which is also located beneath the vertebral column). The veins all gather into one blood vessel before entering the heart. The total volume of the blood in fish ranges from 1.5 to 3.0 per cent of the body weight. Most of it is located in the internal organs while the muscular tissues, constituting two- thirds of the body weight, contain only 20 per cent of the blood volume. This distribution is not changed during exercise since the light muscle in particular is not very vascularized.

During blood circulation the blood pressure drops from around 30 mg Hg in the ventral aorta to 0 when entering the heart . After the blood has passed through the gills, the blood pressure derived from the pumping activity of the heart is already greatly decreased. Muscle contractions are important in pumping the blood back to the heart and counterflow is prevented by a system of paired valves inside the veins. Clearly, the single circulation of fish is fundamentally different from the system in mammals, where the blood passes through the heart twice and is propelled out into the body under high pressure due to the contractions of the heart.

In fish, the heart does not play an important role in the transportation of blood from the capillaries back to the heart. This has been confirmed in an experiment where the impact of different bleeding procedures on the colour of cod fillets was examined. No difference could be found regardless of whether the fish had been bled by means of cutting the throat in front of or behind the heart before gutting, or had not been cut at all before slaughter.

In some fisheries, bleeding of the fish is very important as a uniform white fillet is desirable. In order to obtain this, a number of countries have recommended that fish are bled for a period (15-20 min) prior to being gutted. This means that throat cutting and gutting must be carried out in two separate operations and that special arrangements (bleeding tanks) must be provided on deck. This complicates the working process (two operations instead of one), time-consuming for the fishermen and increases the time-lag before the fish is chilled. Furthermore it requires extra space on an otherwise crowded working deck.

Several researchers have questioned the necessity of handling the fish in a two-step procedure involving a special bleeding period. There seems to be general agreement about the following:

- Bleeding is more affected by time onboard prior to bleeding/gutting than by the actual bleeding/gutting procedure.
- Best bleeding is obtained if live fish are handled, but it is of major importance to cut the fish before it enters rigor mortis since it is the muscle contractions that force the blood out of the tissues.

Disagreement exists as to the cutting method. Huss and Asenjo found best bleeding if a deep throat cut including the dorsal aorta was applied, but this was not confirmed in the work of Botta *et al.* (1986). The latter also recommended to include a bleeding period (two-step procedure) when live fish were handled (fishing with pound net, trap, seine, longline or jigging), while Valdimarsson *et al.* (1984) found that the quality of dead cod (4 h after being brought onboard) was slightly improved using the two-step procedure. However, it should be pointed out that the effect of bleeding should also be weighted against the advantages of having a fast and effective handling procedure resulting in rapid chilling of the catch.

Discoloration of the fillet may also be a result of rough handling during catch and catch handling while the fish is still alive. Physical mishandling in the net (long trawling time, very large catches) or on the deck (fishermen stepping on the fish or throwing boxes, containers and other items on top of the fish) may cause bruises, rupture of blood vessels and blood oozing into the muscle tissue (haematoma). Heavy pressure on dead fish, when the blood is clotted (*e.g.*, overloading of fish boxes) does not cause discoloration, but the fish may suffer a serious weight loss.

Other Organs

Among the other organs, only the roe and liver play a major role as foodstuffs. Their size depends on the fish species and varies with life cycle, feed intake and season. In cod the weight of the roe varies from a few percent up to 27 per cent of the body weight and the weight of the liver ranges from 1 to 4.5 per cent. Likewise, the composition can change and the oil content of the liver vary from 15 to 75 per cent, with the highest values being found during autumn .

GROWTH AND REPRODUCTION

During growth it is the size of each muscle cell that increases rather than the number of muscle cells. Also, the proportion of connective tissue increases with age. Most fish become sexually mature when they reach a size characteristic of the species and is this not necessarily directly correlated with age. In general, this critical size is reached earlier in males than in females. As the growth rate decreases after the fish has reached maturity, it is therefore often an economic advantage to rear female fish in aquaculture. Every year mature fish use energy to build up the gonads (the roe and milk). This gonadal development causes a depletion of the protein and lipid reserves of the fish since it takes place during a period of low or no food intake.

In North Sea cod it was found that prior to spawning the water content of the muscle increases and the protein content decreases. In extreme cases the water content of very large cod can attain 87 per cent of the body weight prior to spawning .

The length of the spawning season varies greatly between species. Most species have a marked seasonal periodicity, while some have ripe ovaries for nearly the whole year. The depletion of the reserves of the fish during gonadal development can be extremely severe, especially if reproduction is combined with migration to the breeding grounds. Some species, *e.g.*, Pacific salmon *(Oncorhynchus* spp.), eel *(Anguilla anguilla)* and others, manage to migrate only once, after which they degenerate and die. This is partly because these species do not eat during migration so that, in the case of a salmon, it can lose up to 92 per cent of its lipid, 72 per cent of its protein and 63 per cent of its ash content during migration and reproduction . On the other hand, other fish species are capable of reconstituting themselves completely after spawning for several years. The North Sea cod lives for about eight years before spawning causes its death, and other species can live even longer . In former times, 25-year-old herring *(Clupea harengus)* were not unusual in the Norwegian Sea, and plaice *(Pleuronectes platessa) up* to 35 years old have been found. One of the oldest fish reported was a sturgeon *(Acipenser sturio)* from Lake Winnebago in Wisconsin. According to the number of rings in the otolith, it was over 100 years old.

FISHING OPERATIONS IN DUTIES OF ALL STATES

States should ensure that only fishing operations allowed by them are conducted within waters under their jurisdiction and that these operations are carried out in a responsible manner. States should maintain a record, updated at regular intervals, on all authorisations to fish issued by them.

States should maintain, in accordance with recognised international standards and practices, statistical data, updated at regular intervals, on all fishing operations allowed by them.

States should, in accordance with international law, within the framework of subregional or regional fisheries management organisations or arrangements, cooperate to establish systems for monitoring, control, surveillance and enforcement of applicable measures with respect to fishing operations and related activities in waters outside their national jurisdiction. States should ensure that health and safety standards are adopted for everyone employed in fishing operations. Such standards should be not less than the minimum requirements of relevant international agreements on conditions of work and service. States should make arrangements individually, together with other States or with the appropriate international organisation to integrate fishing operations into maritime search and rescue systems.

States should enhance through education and training programmes the education and skills of fishers and, where appropriate, their professional qualifications. Such programmes should take into account agreed international standards and guidelines.

States should, as appropriate, maintain records of fishers which should, whenever possible, contain information on their service and qualifications, including certificates of competency, in accordance with their national laws. States should ensure that measures applicable in respect of masters and other officers charged with an offence relating to the operation of fishing vessels should include provisions which may permit, *inter alia*, refusal, withdrawal or suspension of authorisations to serve as masters or officers of a fishing vessel.

States, with the assistance of relevant international organisations, should endeavour to ensure through education and training that all those engaged in fishing operations be given information on the most important provisions of this Code, as well as provisions of relevant international conventions and applicable environmental and other standards that are essential to ensure responsible fishing operations.

MAINTAIN RECORDS OF FISHING VESSELS

Flag States should maintain records of fishing vessels entitled to fly their flag and authorised to be used for fishing and should indicate in such records details of the vessels, their ownership and authorisation to fish. Flag States should ensure that no fishing vessels entitled to fly their flag fish on the high seas or in waters under the jurisdiction of other States unless such vessels have been issued with a Certificate of Registry and have been authorised to fish by the competent authorities. Such vessels should carry on board the Certificate of Registry and their authorisation to fish. Fishing vessels authorised to fish on the high seas or in waters under the jurisdiction of a State other than the flag State, should be marked in accordance with uniform and internationally recognisable vessel marking systems such as the FAO Standard Specifications and Guidelines for Marking and Identification of Fishing Vessels. Fishing gear should be marked in accordance with national legislation in order that the owner of the gear can be identified. Gear marking requirements should take into account uniform and internationally recognisable gear marking systems.

Flag States should ensure compliance with appropriate safety requirements for fishing vessels and fishers in accordance with international conventions, internationally agreed codes of practice and voluntary guidelines. States should adopt appropriate safety requirements for all small vessels not covered by such international conventions, codes of practice or voluntary guidelines. States not party to the Agreement to Promote Compliance with International Conservation and Management Measures by Vessels Fishing in the High Seas should be encouraged to accept the Agreement and to adopt laws and regulations consistent with the provisions of the Agreement.

Flag States should take enforcement measures in respect of fishing vessels entitled to fly their flag which have been found by them to have contravened applicable conservation and management measures, including, where appropriate, making the contravention of such measures an offence under national legislation.

Sanctions applicable in respect of violations should be adequate in severity to be effective in securing compliance and to discourage violations wherever they occur and should deprive offenders of the benefits accruing from their illegal activities. Such sanctions may, for serious violations, include provisions for the refusal, withdrawal or suspension of the authorisation to fish.

Flag States should promote access to insurance coverage by owners and charterers of fishing vessels. Owners or charterers of fishing vessels should carry sufficient insurance cover to protect the crew of such vessels and their interests, to indemnify third parties against loss or damage and to protect their own interests.

Flag States should ensure that crew members are entitled to repatriation, taking account of the principles laid down in the "Repatriation of Seafarers Convention, 1987". In the event of an accident to a fishing vessel or persons on board a fishing vessel, the flag State of the fishing vessel concerned should provide details of the accident to the State of any foreign national on board the vessel involved in the accident. Such information should also, where practicable, be communicated to the International Maritime Organisation.

Port State Duties

Port States should take, through procedures established in their national legislation, in accordance with international law, including applicable international agreements or arrangements, such measures as are necessary to achieve and to assist other States in achieving the objectives of this Code, and should make known to other States details of regulations and measures they have established for this purpose. When taking such measures a port State should not discriminate in form or in fact against the vessels of any other State. Port States should provide such assistance to flag States as is appropriate, in accordance with the national laws of the port State and international law, when a fishing vessel is voluntarily in a port or at an offshore terminal of the port State and the flag State of the vessel requests the port State for assistance in respect of non-compliance with subregional, regional or global conservation and management measures or with internationally agreed minimum standards for the prevention of pollution and for safety, health and conditions of work on board fishing vessels.

Fishing Activities

States should ensure that fishing is conducted with due regard to the safety of human life and the International Maritime Organisation International Regulations for Preventing Collisions at Sea, as well as International Maritime Organisation requirements relating to the organisation of marine traffic, protection of the marine environment and the prevention of damage to or loss of fishing gear. States should prohibit dynamiting, poisoning and other comparable destructive fishing practices. States should make every effort to ensure that documentation with regard to fishing operations, retained catch of fish and non-fish species and, as regards discards, the information required for stock assessment as decided by relevant management bodies, is collected and forwarded systematically to those bodies. States should, as far as possible, establish programmes, such as observer and inspection schemes, in order to promote compliance with applicable measures. States should promote the adoption of appropriate technology, taking into account economic conditions, for the best use and care of the retained catch. States, with relevant groups from industry, should encourage the development and implementation of technologies and operational methods that reduce discards. The use of fishing gear and practices that lead to the discarding of catch should be discouraged and the use of fishing gear and practices that increase survival rates of escaping fish should be promoted.

States should cooperate to develop and apply technologies, materials and operational methods that minimize the loss of fishing gear and the ghost fishing effects of lost or abandoned fishing gear. States should ensure that assessments of the implications of habitat disturbance are carried out prior to the introduction on a commercial scale of new fishing gear, methods and operations to an area. Research on the environmental and social impacts of fishing gear and, in particular, on the impact of such gear on biodiversity and coastal fishing communities should be promoted.

Fishing Gear Selectivity

States should require that fishing gear, methods and practices, to the extent practicable, are sufficiently selective so as to minimize waste, discards, catch of non-target species, both fish and non-fish species, and impacts on associated or dependent species and that the intent of related regulations is not circumvented by technical devices. In this regard, fishers should cooperate in the development of selective fishing gear and methods. States should ensure that information on new developments and requirements is made available to all fishers. In order to improve selectivity, States should, when drawing up their laws and regulations, take into account the range of selective fishing gear, methods and strategies available to the industry. States and relevant institutions should collaborate in developing standard methodologies for research into fishing gear selectivity, fishing methods and strategies. International cooperation should be encouraged with respect to research programmes for fishing gear selectivity, and fishing methods and strategies, dissemination of the results of such research programmes and the transfer of technology.

Protection of the Aquatic Environment

States should introduce and enforce laws and regulations based on the International Convention for the Prevention of Pollution from Ships, 1973, as modified by the Protocol of 1978 relating thereto (MARPOL 73/78). Owners, charterers and managers of fishing vessels should ensure that their vessels are fitted with appropriate equipment as required by MARPOL 73/78 and should consider fitting a shipboard compactor or incinerator to relevant classes of vessels in order to treat garbage and other shipboard wastes generated during the vessel's normal service. Owners, charterers and managers of fishing vessels should minimize the taking aboard of potential garbage through proper provisioning practices. The crew of fishing vessels should be conversant with proper shipboard procedures in order to ensure discharges do not exceed the levels set by MARPOL 73/78. Such procedures should, as a minimum, include the disposal of oily waste and the handling and storage of shipboard garbage.

Energy Optimisation

States should promote the development of appropriate standards and guidelines which would lead to the more efficient use of energy in harvesting and post-harvest activities within the fisheries sector. States should promote the development and transfer of technology in relation to energy optimisation within the fisheries sector and, in particular, encourage owners, charterers and managers of fishing vessels to fit energy optimisation devices to their vessels.

Protection of the Atmosphere

States should adopt relevant standards and guidelines which would include provisions for the reduction of dangerous substances in exhaust gas emissions. Owners, charterers and managers of fishing vessels should ensure that their vessels are fitted with equipment to reduce emissions of ozone depleting substances. The responsible crew members of fishing vessels should be conversant with the proper running and maintenance of machinery on board. Competent authorities should make provision for the phasing out of the use of chlorofluorocarbons (CFCs) and transitional substances such as hydrochlorofluorocarbons (HCFCs) in the refrigeration systems of fishing vessels and should ensure that the shipbuilding industry and those engaged in the fishing industry are informed of and comply with such provisions. Owners or managers of fishing vessels should take appropriate action to refit existing vessels with alternative refrigerants to CFCs and HCFCs and alternatives to Halons in fire fighting installations. Such alternatives should be used in specifications for all new fishing vessels. States and owners, charterers and managers of fishing vessels as well as fishers should follow international guidelines for the disposal of CFCs, HCFCs and Halons.

Harbours and Landing Places for Fishing Vessels

States should take into account, *inter alia*, the following in the design and construction of harbours and landing places:

- Safe havens for fishing vessels and adequate servicing facilities for vessels, vendors and buyers are provided;
- Adequate freshwater supplies and sanitation arrangements should be provided;
- Waste disposal systems should be introduced, including for the disposal of oil, oily water and fishing gear;
- Pollution from fisheries activities and external sources should be minimized; and
- Arrangements should be made to combat the effects of erosion and siltation.

States should establish an institutional framework for the selection or improvement of sites for harbours for fishing vessels which allows for consultation among the authorities responsible for coastal area management.

States should ensure that the standards and guidelines for the removal of redundant offshore structures issued by the International Maritime Organisation are followed. States should also ensure that the competent fisheries authorities are consulted prior to decisions being made on the abandonment of structures and other materials by the relevant authorities.

Fish Aggregation Devices

States, where appropriate, should develop policies for increasing stock populations and enhancing fishing opportunities through the use of artificial structures, placed with due regard to the safety of navigation, on or above the seabed or at the surface.

Research into the use of such structures, including the impacts on living marine resources and the environment, should be promoted. States should ensure that, when selecting the materials to be used in the creation of artificial reefs as well as when selecting the geographical location of such artificial reefs, the provisions of relevant international conventions concerning the environment and safety of navigation are observed.

States should, within the framework of coastal area management plans, establish management systems for artificial reefs and fish aggregation devices. Such management systems should require approval for the construction and deployment of such reefs and devices and should take into account the interests of fishers, including artisanal and subsistence fishers. States should ensure that the authorities responsible for maintaining cartographic records and charts for the purpose of navigation, as well as relevant environmental authorities, are informed prior to the placement or removal of artificial reefs or fish aggregation devices.

Bibliography

A.C. Long: *A-Z of Industrial Marine Fisheries in Asia*, Cyber Tech Publications, Delhi, 2011.

Ajit Kumar Roy and Niranjan Sarangi: *Applied Bioinformatics Statistics and Economics in Fisheries Research*, New India Pub Agency, 2008.

B K Singh: *Applied Fisheries and Aquaculture*, Swastik, Publication, Delhi, 2008.

C.B.L. Srivastava: *A Text Book of Fishery Science and Indian Fisheries*, Kitab Mahal, Delhi, 2013.

D.N. Mishra: *Aquatic Resources and Fisheries*, Surendra Publication, Delhi, 2010.

Gary T. Sakagawa: *Assessment Methodologies and Management: Proceedings of the World Fisheries Congress*, Oxford & IBH, 2007.

H.R. Singh and W.S. Lakra: *Coldwater Aquaculture and Fisheries*, Narendra, Publication, Delhi, 2000.

J.K. Roshan: *A-Z Fish and Fisheries*, Centrum Press, Delhi, 2009.

J.S. Datta Munshi: *Fisheries and Fish Biology Research*, Narendra Publication, Delhi, 2003.

K. Gopakumar and A.D. Diwan: *Coastal Fisheries and Aquaculture Management*, Narendra, Publication, 2010.

K.D. Bhardwaj: *A-Z of Fisheries and Aquaculture Technology*, Cyber Tech Publication, Delhi, 2011.

K.D. Bhardwaj: *Business Management in Fisheries and Aquaculture*, Cyber Tech Publication, 2012.

K.L. Verma: *Global Fisheries Crisis*, Pearl Books, 2012.

K.P. Biswas: *Advancement of Fish Fisheries and Technology*, Narendra Publishing House, 2012.

Neha Charan: *Classification of Fisheries*, Random Publications, Delhi, 2013.

P Ramchandran Nair: *Biotechnology and Genetics in Fisheries and Aquaculture*, Dominant, Publication Delhi, 2008.

P Shambhu: *Biotechnology for Agriculture and Fisheries*, Pearl Books, 2008.

Q. J. Shammi and S. Bhatnagar: *Applied Fisheries*, Agrobios, Publication, Delhi, 2010.

Rajendra Kumar: *A Textbook of Fish and Fisheries*, Arise Publication, Delhi, 2008.

Rajendra Kumar: *Biotechnology and Genetics in Fisheries and Acquaculture*, Arise Publication, Delhi, 2010.

S S Khanna and H R Singh: *A Text Book of Fish Biology and Fisheries*, Narendra Publication, Delhi, 2015.

Syed Hasib Ahmad: *Advances in Fisheries and Fish Production*, Hindustan, Publication, Delhi, 2004.

V P Agrawal: *Aquaculture and Fisheries Science*, S.R. Scientific, Publication, 2006.

Vishwas B Sakhare: *Applied Fisheries*, Daya, Publication, Delhi, 2007.

Yougesh Kumar and Rajeev Tyagi: *Aquaculture Fisheries Biotechnology and Genetics*, Manglam Publishers, 2013.

Index